Environmental Protection Policy and Experience in the U.S. and China's Western Regions

Challenges Facing Chinese Political Development

Series Editor: Sujian Guo, Ph.D.
San Francisco State University

In an attempt to reflect the rapidly changing political environment of the People's Republic of China, editor Sujian Guo has assembled a book series to present specialized areas of research in current Chinese political studies. Incorporating theoretical, empirical, and policy research on contemporary Chinese politics both domestically and internationally, this series contemplates the Chinese past, present, and future by utilizing interdisciplinary perspectives to approach issues related to Chinese politics, economy, culture, social development, reform, the military, legal system, and foreign relations. Aimed at bringing a greater understanding of the current Chinese political climate to Western audiences, this series is focused on the emerging voices of Chinese scholars and their perspectives on the ever-changing Chinese diaspora.

Recent titles in the series are:

The Dragon's Hidden Wings: How China Rises with Its Soft Power, by Sheng Ding

"Harmonious World" and China's New Foreign Policy, by Sujian Guo and Jean-Marc F. Blanchard

China in Search of a Harmonious Society, edited by Sujian Guo and Baogang Guo

Greater China in an Era of Globalization, edited by Sujian Guo and Baogang Guo

Toward Better Governance in China, edited by Baogang Guo and Dennis Hickey

Dynamics of Local Governance in China during the Reform Era, edited by Tse-Kang Leng and Yun-han Chu

Dancing with the Dragon: China's Emergence in the Developing World, edited by Dennis Hickey and Baogang Guo

Online Chinese Nationalism and China's Bilateral Relations, edited by Simon Shen and Shaun Breslin

Multidimensional Diplomacy of Contemporary China, edited by Simon Shen and Jean-Marc F. Blanchard

Thirty Years of China-U.S. Relations: Analytical Approaches and Contemporary Issues, edited by Sujian Guo and Baogang Guo

Environmental Protection Policy and Experience in the U.S. and China's Western Regions, edited by Sujian Guo, Joel J. Kassiola, and Zhang Jijiao

Environmental Protection Policy and Experience in the U.S. and China's Western Regions

Edited by Sujian Guo,
Joel J. Kassiola, and Zhang Jijiao

LEXINGTON BOOKS
A division of
ROWMAN & LITTLEFIELD PUBLISHERS, INC.
Lanham • Boulder • New York • Toronto • Plymouth, UK

Published by Lexington Books
A division of Rowman & Littlefield Publishers, Inc.
A wholly owned subsidiary of The Rowman & Littlefield Publishing Group, Inc.
4501 Forbes Boulevard, Suite 200, Lanham, Maryland 20706
http://www.lexingtonbooks.com

Estover Road, Plymouth PL6 7PY, United Kingdom

British Library Cataloguing in Publication Information Available

Library of Congress Cataloging-in-Publication Data
Environmental protection policy and experience in the U.S. and China's western regions / edited by Sujian Guo, Joel J. Kassiola and Zhang Jijiao.
p. cm. — (Challenges facing Chinese political development)
Includes bibliographical references and index.
ISBN 978-0-7391-4742-9 (cloth : alk. paper)
1. Environmental policy—United States. 2. Environmental protection—United States. 3. Environmental policy—China. 4. Environmental protection—China. 5. United States—Environmental conditions. 6. United States—Politics and government. 7. China—Environmental conditions. 8. China—Politics and government. I. Guo, Sujian, 1957- II. Kassiola, Joel Jay, 1945- III. Zhang, Jijiao.
GE180.E637 2010
363.7'05610951—dc22
2010009318

The paper used in this publication meets the minimum requirements of American National Standard for Information Sciences—Permanence of Paper for Printed Library Materials, ANSI/NISO Z39.48-1992.

Printed in the United States of America

Contents

List of Figures

List of Tables

Contributors

Fang Sumei is Professor, Senior Research Fellow, and Vice Head of the Department of Ethno-History at the Institute of Ethnology and Anthropology, Chinese Academy of Social Sciences. Her research interests are focused on the history of China's minority nationalities, ethnic issues, and ethnic policies in modern China, and social and cultural changes experienced by minority nationalities in modern China. Her authored and edited books include *A Study of Zhuang Society in Modern Times* (Nanning, Guangxi Ethnic Publishing House, 2002); *A History of the Minority Revolution in China, 1840-1949* (Nanning, Guangxi Ethnic Publishing House, 2000); *Disasters, Culture, Politics: Chinese-Bulgarian Anthropological Contribution to the Study of Critical Situations* (Newcastle, Cambridge Scholars Publishing, 2009).

Sheldon Gen is Assistant Professor of Public Administration at San Francisco State University. He studies the human behavioral aspects of public policy, including how citizens participate in policy making and how they respond to public policies. His research has appeared in *Environmental Practice*, *Environment & Behavior*, *Sustainable Development*, *Health and Place*, and in other journals pertaining to public administration and policy. Prior to his academic career, he was a civil and environmental engineer with several American government agencies including the U.S. Environmental Protection Agency, the U.S. Peace Corps, the U.S. Air Force, and state and local agencies. He is also an alumnus of the Presidential Management Fellowship in the United States. He holds a B.S. in civil engineering from the California Polytechnic State University, a license in civil engineering in the State of California, an M.P.A. in public administration from the University of Southern California, and a Ph.D. in public policy from the Georgia Institute of Technology.

Qian Guo is Associate Professor in the Department of Geography and Human Environmental Studies at San Francisco State University. He received his Ph.D. from the University of Tennessee, Knoxville. His research interests have focused on frontier development and agricultural conservation, with regional focus on the Chinese western regions and the American West.

Sujian Guo is Professor in the Department of Political Science and Director of Center for U.S.-China Policy Studies at San Francisco State University. He holds concurrent appointments as Associate Dean and Distinguished Guest Professor of Fudan Institute for Advanced Study in Social Sciences at Fudan University. He is Editor-in-Chief of the *Journal of Chinese Political Science*, Associate Editor-in-Chief of the *Fudan Journal of the Humanities and Social Sciences,* Series Editor of Rowman & Littlefield-Lexington's Chinese political development, and former President of the Association of Chinese Political Studies. He received his MA degree from Peking University and Ph.D. from the University of Tennessee. His research interests include Chinese/Asian politics, U.S.-China relations, communist and post-communist studies, democratic transitions, and the political economy of East and Southeast Asia. He has published more than 30 academic articles both in English and Chinese. His authored and edited books include *Thirty Years of China-U.S. Relations: Analytical Approaches and Contemporary Issues* (2010);*Greater China in an Era of Globalization* (2009); *Harmonious World and Chinese New Foreign Policy* (2008); *China in Search of a Harmonious Society* (2008); *Challenges Facing Chinese Political Development* (2007); *New Dimensions of Chinese Foreign Policy* (2007); *China in the Twenty-First Century: Challenges and Opportunities* (2007); *The Political Economy of Asian Transition from Communism* (2006); *China's 'Peaceful Rise' in the 21st Century: Domestic and International Conditions* (2006); and *Post-Mao China: From Totalitarianism to Authoritarianism*? (2000).

Shiyuan Hao is Director and Academician in the Division of Social Politics and Law in Chinese Academy of Social Sciences (CASS), Director and Professor in the Institute of Ethnology and Anthropology of CASS, Director of Oversea Chinese Research Center of CASS, and President of Chinese Association for Ethnology. He received his MA from the Graduate School of CASS. His research interests include environment and sustainability, multiculturalism and ethnic studies. In the past few years he paid more attention to environmental issues of China's Western Regions. He has published more than 80 academic articles both in English and Chinese. His published articles include "Toward a Socialist Harmonious Society: Embracing and Respecting Diversity" (2007), "Respect Differences and Tolerate Diversities: Social Inclusion in Multi-ethnic Countries" (2006), "Ecological Migration in China" (2005), and "Building a Harmonious Society and Fastening the Development of Western Region" (2003).

Barbara A. Holzman is Professor in the Department of Geography and Human Environmental Studies and founding director of the Environmental Studies Program at San Francisco State University. She received her Ph.D. from the University of California at Berkeley. Her research interests are the ecology and politics of natural resource management, rangeland and forest ecology, and tropical rainforest biogeography. She is author of several books and articles including a recent book on Tropical Forest Biomes published by Greenwood Publishing in 2008.

Joel J. Kassiola is Dean of the College of Behavioral and Social Sciences and Professor of Political Science at San Francisco State University. He received his M.A. and Ph.D. degrees from the Political Philosophy Program at Princeton

University. His research program after several publications in various political theory themes: the conduct of normative inquiry, politics and literature, the justification for affirmative action, political violence, turned to the intersection of the environmental crisis, political theory and modernity about 30 years ago. He published one of the first books in the emerging field of environmental political theory in 1990: *The Death of Industrial Civilization* (SUNY Press), and more recently, edited *Explorations in Environmental Political Theory* (M.E. Sharpe, 2003). He has published several articles on the nature of environmental political theory and environmental ethics. Most recently, he has published an article in the flagship political science journal published by the American Political Science Association on the teaching of Introductory Political Theory courses, and article on the need for M.A. Universities to have their faculty conduct research in order to be excellent teachers in the 21st century. His most recent research has turned to China's political thought and development pertaining to green political theory, environmental policy, and China's development. His article titled "The Dilemma of Western Industrial Civilization and China's Path in the 21st Century" is published in *Challenges Facing Chinese Political Development* (Rowman & Littlefield-Lexington, 2007). He is an editor and a contributor of a book titled *China's Environmental Crisis: Domestic and Global Political Impacts and Responses* to be published by Palgrave MacMillan in 2010.

Li Yujun is Associate Professor and Director at the Department of Environmental Economics and Management in the Institute of Urban and Environmental Studies, Chinese Academy of Social Sciences. She received her MA degree from Peking University. Her research interests include environmental policy, sustainable development, municipal solid waste management, environmental impact assessment, environmental standards. She was in charge of over 20 research projects on environmental issues. One research project on environmental standard, "State Discharge Standard of Pollutants for Municipal Wastewater Treatment Plant," won the Third Prize of Environmental Science and Technology of Ministry of Environmental Protection of the People's Republic of China in 2004. Her most recently published papers include "Urban Environmental Sustainable Development in China," "Urban Environment Treatment in China," "The Evaluation of Chinese Public Environmental Quality: A Comparative Study Based on Five Surveys on Public Environmental Quality Since 1999," "Research on the Improvement of Municipal Solid Waste Management in China," and "Optimization of Urban Functions and Livable City."

Xiao Hang Liu is Associate Professor in the Department of Geography and Human Environmental Studies at San Francisco State University. She received her Bachelor's degree in Environmental Science from Peking University in China, her Master's degree in Geography and Computer Science from Rutgers University, New Brunswick, N.J., and her Ph.D. in Geography from University of California, Santa Barbara. Liu's teaching and research focus on Geographical information Science and urban remote sensing. Her research usually employs high resolution satellite images to detect urban changes, then simulates the change process by computer modeling. She is the author of nearly 10 peer-reviewed journal articles, and has published with *International Journal of Geographical Information Science, Photogrammetric Engineering and Remote Sensing*, etc.

Gary W. Pahl is Professor of Archaeology with the Department of Anthropology at San Francisco State University. He has previously been chair of the department for eleven years. He received his M.A. in Archaeology and Ph.D. in Anthropology from U.C.L.A., focusing on the ancient Maya inscriptions at Copan, Honduras. As a post-doctoral researcher he served as Executive Editor for the U.C.L.A. Latin American Center for two years. His research interests, field work experience and publications have ranged from prehistoric California site research in the San Francisco Bay Area to ancient Chinese archaeological research on Late Neolithic walled towns in Hubei and Sichuan Provinces with Professor He Nu of the Chinese Academy of Social Sciences in Beijing. In Maya Archaeology he has conducted excavations related to Copan at the Maya sites of La Canteada and Rio Amarillo and surrounding sites. In Guatemala he conducted research and excavation at the ceremonial center of Polol in the Peten region. An underlying focus of his research in China and the Maya region is the study of ancient writing systems from both areas. Representative publications in English, Spanish and Chinese across these areas include *The Periphery of the Classic Maya Realm* (1987), *The Archaeology of De Silva Island, Mrn-17, California* (2003), "Shang Kingdom Terrorism in Analytical Comparison with Aztecs of Mexico" (2005), and An English-Chinese Dictionary for Chinese Archaeology in preparation.

Yin Peng received his Master's degree in urban anthropology from the Graduate School of Chinese Academy of Social Sciences, and he is a researcher at Shanghai Academy of Social Sciences since 2008.

Guoqing Shi is Professor and Dean of Public Administration School of Hohai University, Director of National Research Centre for Resettlement (NRCR), Director of Social Development Institute of Hohai University, Chair of Membership Committee of International Water Resource Association (IWRA), Advisor of Three Gorges Dam Project, Consultant of World Bank and Asian Development Bank, and Vice Editor-in-Chief of the *Journal of Water Science and Engineering*. He received his BA and MA degrees from Hohai University. His research interests include population migration and resettlement, social and economic assessment, environment and resource administration, and management. He has published more than 180 academic articles both in English and Chinese. His authored and edited books include *China Resettlement Policy and Practice* (2001), *Reservoir Resettlement System Theory and Application* (1996), *Resettlement and Social Development* (2002), and *Social Integration of Migrants* (2004). Translated books include *Resettlement and Development—the World Bank Resettlement Policies and Experience* (1996), *Resettlement, Reconstruc-*

tion and Development—the World Bank Resettlement Policies and Experience (1998).

Guan Yi received her Master's degree in urban anthropology from the Graduate School of Chinese Academy of Social Sciences, and she is a researcher at China Academy of Arts Research since 2008.

Zhang Jijiao is Professor in the Institute of Ethnology and Anthropology, Chinese Academy of Social Sciences. He is Founder and Chair of the Commission on Enterprise Anthropology, IUAES, since 2009. He is also Assistant Secretary of the Commission on Urban Anthropology, IUAES, 2004-2006, Vice Secretary-in-General, Organizing Committee for the 16th IUAES World Congress, 2003-2009; and Vice Secretary-in-General of the China Urban Anthropology Association, 1998-2003. His research areas include urban anthropology, enterprise anthropology and urban environment studies. He has published six papers and one book in English and Japanese.

Qiu Zhong is a Ph.D. candidate in the Department of Public Administration at Hohai University, visiting graduate student in Science and Art Department at Oklahoma State University (2008-2009). She received her bachelor degree from Nanjing Agriculture University and MA degree from Hohai University. Her research interests include environmental consciousness, social culture and nature environment, and resettlement adaptation. She has published three academic articles in Chinese: "Administration of Water Resources in China" (2006); "Face and Regulation in China's Rural Area" (2006); and "The Review of History: China Environmental Events" (2009).

Jian Zhou is Assistant Professor of Hohai University, Social and Resettlement consultant of World Bank and Asian Development Bank, Member of International Association for the Study of Forced Migration and Association of Social Science of China. He received his Ph.D degree from Hohai University. His research interests include involuntary resettlement induced by development projects, environmental sociology, social impact assessment and engineering sociology. He has published more than ten academic articles both in English and Chinese, and his most recent publications include "Analysis on Policy and Effect of Ecological Migration" (2009), "Optimization Selection for Relocation Areas of Ecological Migration Based on Fuzzy Logic" (2009), "Study on Social Assessment for Urban Water Environment Improvement Project" (2008), "Case Study of Ecological Migration along Luntai Tarim River Nature Reserve" (2008), and "A Cost-Benefit Analysis of Rural Labor Migration" (2006).

Acknowledgments

This book is a collection of papers from an international conference titled, "U.S. and China's Environmental Protection Policy: the Impact of National and Local Policy on the Western Region," from March 24–25, 2009, at the Chinese Academy of Social Science's Institute of Ethnology and Anthropology. The conference was co-sponsored and co-organized by San Francisco State University's College of Behavioral and Social Sciences and Center for U.S.-China Policy Studies. The authors would like to thank the above organizations for their generous funding for the conference.

Introduction

Ecological and Environmental Challenges in China's Western Regions

Shiyuan Hao and Joel J. Kassiola

China is a multi-ethnic country with vast territory. The population of ethnic minorities is 11 million, about 9 percent of the nation's total population. However, these regions account for nearly 64 percent of China's total land. Meanwhile, geographically, they are concentrated in western China, which is economically and socially underdeveloped. There is still a deep economic and cultural gap between ethnic minority areas and regions inhabited mainly by Han people, especially the southeastern coastal areas. Therefore, the Chinese government aims to accelerate the cultural and economic development in minority regions in Western China and promote common prosperity of all nationalities. The strategy of "Western Grand Development," initiated at the beginning of the new century, is one of the major measures to achieve these basic goals, and it constitutes a major part of China's modernization. Although western China is economically and socially underdeveloped, it has the richest resources in its cultural and biological diversity. Therefore, western development is not only the development of economy, but also a common challenge to contemporary human society—what we should do when development encounters cultural, civilizational and ecological issues.

China, a land of diverse ecosystems, ranks the third in the world concerning her abundance of plant resources with over half of global plants unique to China. At the same time, China, with its mammals, birds, and fish resources accounting for one-eighth of the world, is one of the global centers for genetic variation of crop cultivation. Therefore, sustainable use of biodiversity resources in China has been facing a great threat because of its enormous population size and the growing demand for natural resources that has accompanied rapid economic and social development since reform and opening. Especially in the process of economic and social development, China's ecological environmental problems have become increasingly prominent as a result of growing interaction between human activities and natural environment as well as global warming.

In the "Western Grand Development," ecological balance and preservation should be the heart of the western development strategy through biodiversity conservation measures based on natural reserves. To meet the challenges, China must adopt a series of important measures not only to reduce industrial emissions, but also return farmland to forests and pasture to grasslands conducive to the cultivation of the ecological environment, among which ecological migration is required to reduce human activities in ecological conservation areas. In this regard, China must not only learn profound lessons from industrialized countries but also search for international cooperation.

It is in this context that an international conference titled "Environmental Protection Policy: The Impact of National and Local Policy on the Western Region—U.S. and China Experience" was held from March 24-25 in Beijing, co-organized by San Francisco State University's Center for U.S.-China Policy Studies, College of Behavioral and Social Sciences, and the Chinese Academy of Social Science's Institute of Ethnology and Anthropology. This edited book is based on the selected papers of the conference. The contributors to this volume are environmental scholars and specialists from the U.S. and China. Their insightful comparisons and analyses would help students of this academic field understand the biological and environmental challenges facing China in the western regions, their significant global implications, and responses and strategies China should adopt in coping with the biological and environmental problems in the western development. The observational studies conducted by Chinese scholars based on their field trips in western regions are unique in analyzing the state of environmental circumstances in western regions, and should provide first-hand raw material, which is not accessible to "outsiders" to the regions, as well as empirical analyses for further scholarly research on the problems raised by authors.

This book is organized into two parts, and each is composed of five chapters. Part 1 studies the lessons from the U.S. experience for China's environmental theories and practices. Part 2 focuses on the Chinese environmental policy and experience in the Western regions.

In chapter 1, "American Agricultural Conservation as a National Policy: Lessons for China's Great West Development," Qian Guo examines the environment-resource impact of U.S. agricultural policies on the American West and applicability of U.S. agricultural conservation policy to China's Great West Development. Both the United States and China have emphasized agriculture in developing their western regions which had been vast and underdeveloped with fragile ecological equilibriums. While China launched its West Development a decade ago, the United States underwent rapid westward expansion more than a century earlier; lessons can be drawn from western U.S. for China's ongoing endeavors. Until recently, the U.S. federal government appeared to favor production over conservation in its agricultural policy and unsustainable, industrialized agricultural operations have led to serious resource degradation and environmental deterioration. There has been a major shift toward greater emphasis on conservation in federal agricultural policies, but implementation of the new, conservation-oriented policies may be difficult due to both privatization of public resources and entrenchment of the agro-business community. Agricultural conservation is now carried out mainly by state and local governments; private participation has to be coerced with legal instruments and financial incentives. The paper contends that China must formulate strong and enforceable national policies for agricultural conservation from the beginning of its West Development in order to strike a balance between need for short-term, significant production increase and long-term, sustainable practices in the western regions.

In chapter 2, "Grassroots Environmental Management in the Western United States: Can It Be a Model for Success for Western China" Barbara A. Holzman and Qian Guo examine the evolution of grassroots environmental management and its effects on conservation and sustainable development in the western United States. Local interests and involvement in environmental protection have always been strong in western U.S., where conflicts of interests arose among frontier settlers, early environmentalists, and federal and local governments. The latest development in environmental policies and their implementation in the western U.S. is grassroots environmental management. It is a systematic approach focusing on the local level in implementing goals of environmental protection and enabling citizens to participate in natural resources management. It shifts environmental efforts from confrontation to cooperation among the stakeholders, such as local community, resource users, and environmentalists, as well as local, state, and federal agencies. The outcome of this approach is informed environmental policies and better compliance to management strategies. While political and social contexts between western China and the western U.S. are different, the two regions face comparable environmental challenges. Grassroots environmental management may provide a feasible model for western China at a time when China is reorienting itself toward parallel tracks of property rights and sustainable development.

In Chapter 3, "Values Conflict in American Environmental Federalism, with Implications for China," Sheldon Gen asks such a question: as China pursues the economic development of its western regions, can the experiences of the U.S. help it avoid the environmental problems Americans experienced? To answer this question, this chapter focuses on the tensions in American environmental policies in a federal structure of government, and the value conflict reflected in these policies. It first reviews the development of American environmental policy, from western expansion of the 1800s to the contemporary environmental movement. It then reveals the evolution of environmental values reflected in this history, its conflicts with the dominant values of individualism, and the current state of that conflict. Next, it examines how the federal structure of government in the U.S. exacerbates these value conflicts, highlighting example policies and national and local levels that take two sides in the values conflict. Finally, it argues that China is increasingly facing a similar values conflict in its environmental protection efforts, and identifies how it might avoid the problems seen in the American experience.

In chapter 4, "Urban Growth Control and Farmland Preservation in West China," Xiao Hang Liu argues that in order to speed up the industrialization in west China so as to absorb the surplus labors in rural areas, urbanization has been utilized as a key. Current urbanization in west China has mainly taken place in the form of urban sprawl where existing large cities and their suburbs spread themselves over the rural land at the fringe. Geographically, existing large cities in west China are mainly located with fertile soil and easy water access, e.g. diluvian fans, alluvial plains, or oasis along the feet of mountains and rivers. These areas happen to be where prime farmland of high productivity is found. Hence, urban sprawl causes rapid loss of prime farmland which is precious to the arid west and China in general. How to preserve farmland in the face of accelerated urbanization is thus a serious challenge to the decision makers. In this chapter, the author first reviews the environmental problems associated with prime farmland loss in west China, then introduces the policy instruments employed by the U.S. to address similar challenges and lessons learned; from there a few suggestions are provided to China.

In chapter 5, "Archaeological Site Management as Environmental Policy in the United States and China: Implications for Western Expansion in China," Gary W. Pahl argues that China and the U.S. have independently worked to preserve their heritage manifested in archaeological resources and arrived at parallel but unique practices to effect protective control for their cultural resources. This chapter offers a comparative look at how these distinct policies have developed and potentially reveals directions China might contemplate as it experiences a growth surge into its western regions. The chapter draws together the historical sequence of policies in the American experience with cultural resource management dating to the 1880s and its uneven legal modifications in response to changing national and local interests. America's conflicting issues down to the unsolved questions surrounding legal ownership of heritage resources are summarized. In comparison, China's current rigorous federal top-down approach to cultural resource management is seen as effective and might best avoid proliferation of the multiple American layers of conflicting local, state, and national laws which cloud clear solutions.

Chapter 6 is "Environmental Consciousness Change: A Comparative Study of the United States and China (1950-2008)" by Qiu Zhong and Guoqing Shi. This chapter compares the change of environmental consciousness from 1950 to 2008 in the United States and China. Two important factors, science and media, are asserted. However, when environmental consciousness becomes established, the author places emphasis on economics and political factors. When one reviews the developing progress and analyzes the links among the science, media, economic and political factors, one can see the interwoven change of those relations. The United States is at the deepening stage while China has just come into the threshold; this comparative work will help China learn from some of the experiences of the United States.

Chapter 7 is Li Yujun's "Urban Environmental Protection Policies and their Effects in China." Li reviews the process of the environmental protection policy-making and the effects of urban environmental protection in China during the thirty years of the reform and opening-up period, evaluates the important guidelines and measures of the environmental protections in China throughout different historical periods, and analyzes the achievements of the urban environmental protection and the challenges it faced. The chapter concludes by submitting some constructive countermeasures to the challenges and problems.

In chapter 8, "Ecological and Environmental Migration in China: Policy and Practice," Guoqing Shi and Jian Zhou argue that the essence of ecological and environmental migration is to solve ecological and environmental problems through human migration and relocation of economic activities. Urbanization and industrialization have created many real and potential environmental problems that seriously affect the local people's economic activities living conditions, and sometimes even their lives and properties. Environmental migration has emerged as a way of protecting the ecological system, improving the quality of the environment, and protecting the citizens' fundamental rights such as rights to life and health. The typical measures of environmental migration include relocating the polluting factories and the factories which are being polluted with help from the government, society and the polluting factories. In China's western areas, ecological migration organized and implemented by the government to protect the ecological environment has recently become a trend. Based on the above definition of ecological and environmental migration, this chapter analyzes the similarities and differences between ecological and environmental migration and other types of migration. Three causes of environmental migration are also summarized, which include environmental pollution, environmental disasters, and ecological deterioration. Then, the chapter discusses the different methods of migration, which include relocation of polluting sources, removal of the population affected, governance on treatment and ecological migration according to the environmental impacts and the feasibility of implementation. Finally, authors put forward several issues which require further research from sociology, management and economics perspectives.

In chapter 9, "'Cities Besieged by Garbage' and Countermeasures for Environmental Protection in Western Cities—A Case Study of Solid Waste Classification in Kunming," Zhang Jijiao argues that "cities besieged by garbage" have become one of the most urgent and important problems for urban environmental protection. Recently, environmentalists and experts on garbage disposal have been dealing with the urban solid waste issue from several perspectives—its harms, disposal techniques and rubbish dump construction. However, this issue has never been studied from the perspective of urban residents. Through a case study of Kunming, a large city in western China and the capital city of Yunnan Province, based on questionnaire survey of 215 samples (80 staff and 135 non-staff), this research analyzes interviewees' awareness of solid waste issues and garbage classification, their understanding about government regulations on solid waste administration, behaviors of solid waste classification, attitudes towards separate collection of solid waste and their evaluation and suggestions on the solid waste issue in Kunming. Finally it puts forward some suggestions to tackle solid waste problems: (1) regulations and legislations on solid waste classification should be improved and economic incentives should be established; (2) a system should be built to garbage discharge at fixed times and places; (3) mechanisms should be set up to promote solid waste classification information; (4) pilots test for solid waste classification should be accelerated.

The tenth and final chapter is Fang Sumei's "Ecological Migration Practices and the Placement of Migrants and Its Impact on Poverty Alleviation and Development—A Case Study of the Huanjiang Maonan Autonomous County Guangxi." Fang argues that ecological migration is an important part of environmental policies that exist in many countries. The Chinese government combines immigration placement policies with poverty alleviation and development programs to eliminate poverty in China. In the 1980s, the Huanjiang local government begun to resettle impoverished farmers from stony mountain areas to plain and hilly lands. In addition, between 1993 and 2002, about 60,000 impoverished farmers migrated to Huanjiang from other counties. As a result, Huanjiang has the largest migrant population in Guangxi Zhuang Autonomous Region. The practice of migrant resettlement to alleviate poverty and promote development in Huanjiang reflects ecological migration policy, and it is typical in contemporary China. Therefore, the study of ecological migration in Huanjiang would help us have a better understanding of the Chinese immigration placement policies, actions, experience, results, and lessons for the Chinese government environmental protection policy.

Chapter 1

American Agricultural Conservation as a National Policy: Lessons for China's Great West Development

Qian Guo

A crucial step in China's global emergence is to develop its vast western regions,[1] or the Chinese West. Dubbed "Great West Development" (*xibu da kaifa*), the Chinese government officially launched the national campaign in 1999.[2] Agricultural development takes center stage in China's Great West Development, because of its paramount importance to national food security, resource exploration and sustainable economic growth. It is imperative that China formulate agricultural policies consistent with the overarching goals of Great West Development and national strategies for sustainable development.

China must look beyond its past and borders for reference to its Great West Development in the age of modernity and globalization. The United States, renowned for its western development, may provide both inspiration and cautionary lessons for China. This is especially true in terms of its federal policies regarding western agriculture.

While the United States is a young country in terms of the history of civilization, it has carried out rapid development in its western regions guided by far-reaching national policies. Some of the policies have had direct impact on agricultural development. As a modern state, on the other hand, China is for the first time setting national policies for its vast western regions. What can China learn from the development of the American West,[3] especially regarding government agriculture policies for the western regions?

THE CONQUEST AND DEVELOPMENT OF THE AMERICAN WEST

The geographic imagination and reality of the West have evolved for both the United States and China. For both, the West denoted a historical and geographical frontier, a domain both rich in resources and crucial to national development. Agriculture has been and continues to be the foundation in the settlement and development of the West by providing basic sustenance that retains and sustains settlers. National government has played a vital role in the West for both countries. There are major differences, however, between the two countries in terms of political and economic dynamics. The United States conquered its western frontier swiftly, driven by free enterprise and market demand. It has also had by far a most systematic and legislatively defined approach to agriculture policies. In China, these socioeconomic and political conditions have been weak or entirely absent. As China emerges as a political and economic power, its vast West has become a center stage for new development with debates raging on over the most appropriate. Meanwhile, large-scale development projects have been underway in the West. There were nearly one hundred major projects between 2000 and 2007, with a total investment of 1.3 trillion *yuan* (nearly US$200 billion in the year of 2009).[4] How should the Chinese government manage such rapid and large-scale development? What kind of agriculture development should it promote in its western region? China may be able to take a page out of the U. S. experience in its search for sustainable development policies: adopting the strategies that have made the American West key to its national economic success, and avoiding the mistakes that damaged ecological equilibrium and hindered long-term sustainable development. In this sense, China may want to pay special attention to the traps of industrialized, export-base commercial agriculture that has dominated the American West.

The United States set agriculture as its national foundation from the beginning. George Washington proposed to Congress that a National Board of Agriculture be established to coordinate agricultural development (1799). U. S. Congress established Agricultural Committees in the House (1820) and Senate (1825) respectively. As a young country settled mostly by European immigrants, the United States looked to the west for both imagination and resources in its nation-building efforts. Agricultural settlement led the way of westward movement and settlement. While the U. S. federal government did not have specific policies encouraging commercial agriculture development before the mid 1800s, it did provide security and interstate transportation infrastructure for large-scale westward movement. The state and local government, at the same time, were more directly involved in assisting commercial agricultural activities, such as price regulations, scientific research and product inspection.[5]

Western frontier settlement was crucial to early national development of the United States. The U. S. government encouraged its citizens to move across the Appalachian Mountains and settle the interior as farmers immediately after Independence. The Prairie regions to the west of the Appalachians, known as the Middle West (or Midwest), became developed quickly through agricultural settlement characterized by its commercial nature and export dependency. Agricultural products from the Midwest were mostly shipped to the East Coast and Europe. While the American folklore would let one believe that the frontier settlers were fiercely independent and resisted government intervention, the reality was that they actually received plenty of government help, directly or indirectly. Although there was a lack of national policies regarding western development, the government did play a vital role in assisting westward expansion and western development. In addition to military protection, the most important public support was through the construction of a transportation infrastructure. The Cumberland Pike (1818) and the Erie Canal (1825), as well as other transportation networks, were built by state and local governments to assist the development of new, rapidly expanding commercial agricultural communities.

As settlement moved further west into the lands mostly controlled by the federal government, it became imperative that the United States establish national policies for appropriating public land for private settlement. Without direct government intervention, land distribution would be chaotic and inefficient. The United States government began to take a much more direct role in promoting agricultural settlement and development in its western regions. Three agriculturally-related legislations were enacted in 1862 alone and became the cornerstones for the development of the American West. First, U. S. Congress passed the Homestead Act of 1862,[6] which would grant settlers 160 acres of federally owned land for free if certain conditions were met.

The second legislation passed in 1862 related to agriculture was the Morrill Land Grant Act,[7] which provided each eligible state with 30,000 acres of federal land. The proceeds from land sales funded the establishment of colleges and universities that would emphasize the research and promotion of advanced agricultural technology. Early land grant universities included Massachusetts Institute of Technology (MIT) and Cornell University. Ensuing legislations established more land grant universities and colleges to a total of 106 in 2000.[8] Together, they propelled the United States to the forefront of science and technology and have benefitted American agriculture in the ensuing decades.

A third Congressional legislation that also became law in 1862 was the Pacific Railroad Act.[9] The federal government provided railroad companies with generous land grants as a form of financial aid. While this Act was not directly aimed at agricultural development in the West, the railroad, more than any other forms of transportation, opened up large tracts of western land for settlement and agriculture, and made it possible for agricultural products to be shipped from the West to the East Coast and Midwest markets. Furthermore, railroad companies sold a large portion of their land grants to (agricultural) settlers, providing another form of western land allocation for development.

1862 was perhaps the most important year in U. S. agricultural history. In addition to the three pieces of legislation discussed above, the United States Department of Agriculture (USDA) was founded and subsequently gained cabinet status in 1889. The founding of this new federal Department during the era of limited federal power was an indication that the public recognized the necessity of enforcing laws and managing appropriation of resources at the national level in western and agricultural development. It is important to point out that, in 1862, the United States was in the midst of a devastating Civil War, and the national government was charged with managing the immediate crisis while keeping an eye on long-term national interests. The timing for the founding of USDA was not, therefore, coincidental; it was the result of the country's western imperative and the role agricultural development played in fulfilling such an imperative.

What is the relevance of the early development of the American West to China's Great West Development? There are major differences but also some parallels in terms of challenges and strategies between the American West of the 19th century and the Chinese West of the twenty-first century. First, the Chinese West has long been settled by various groups and nearly all developed with ecologically viable settlement and economic systems based on irrigation agriculture. Any large-scale, new development must respect the integrity of the time-tested traditional way of life for environmental health and social stability. Ecological conditions in Chinese West are even more challenging than those in the American West in terms of both topography and aridity, as well as in strains wrought upon by long-term, exhaustive human exploration. Without significant water supply increase, American-styled large-scale migration and settlement are not feasible strategies for the Chinese West. In addition, the United States relied almost solely on free enterprise and a market system to encourage western development, as idealized in the Turner Thesis.[10] The socioeconomic context is quite different for China's Great West Development, and the *Sooners*-style land reclamation is not possible in the twenty-first century.

Notwithstanding these differences, there are several important parallels between the American West and the Chinese West that may make the U. S. experience transferable. First, the national government's leadership is paramount to regional development, especially the frontier, and government policies would have the greatest impact on whether the subsequent development is sustainable in the long-run. China should endeavor to develop policies for western development with a balance between public sector and the private sector, and facilitate private involvement in Great West Development. Second, infrastructure building provided major impetus for population settlement and agricultural development in the American West. This is being mirrored by the Chinese central government's strategies for its "Great West Development," i.e., with the first priority being "acceleration of infrastructure building." While the first four large-scale public projects prioritized by the government have mainly to do with resource exploration,[11] these projects are to have direct impact on the extent and patterns of agricultural development.

As a result of key federal legislation, large-scale commercial agriculture spearheaded large-scale westward expansion in the United States. The Homestead Act and, to a lesser extent, the Pacific Railroad Act were the greatest and most direct driving forces for rapid settlement, especially that of agriculture, in the American West. Between 1862 and 1986, when the Act was finally suspended in Alaska, 1.6 million homesteads were claimed; over 270,000,000 acres (1.1 million square kilometers), or about 10 percent of total U. S. lands, were granted to private parties. Few other national policies have had such far reaching, and perhaps most controversial, impacts on the development of the United States. The U. S. policies for western development have had desired effects because they were framed in the contexts of its socioeconomic conditions and political configurations. These policies also combined individual interests and national priorities. Furthermore, emphases on extractive activities were supported by resource availability of the American West. Will China's national policies have comparable impact on its western development? It depends upon whether these policies are formulated within the socioeconomic and cultural contexts of the Chinese West and based on visions that can translate into long-term; sustainable planning.

By the end of 1800s, large tracts of "free" or inexpensive land in the West were taken and the U. S. Census Bureau declared that American frontier was closed. The engine set in force by the Homestead Act and Morrill Land Grant Act, however, continued to drive American agricultural development. An increase in capital input was made possible with bank loans for agricultural operators and industrial investment. Technology diffusion was quick and smooth among farmers and ranchers, thanks to an extensive and effective network of land grant universities, USDA extension services (established through the Smith-Lever Extension Act of 1914), and private industries that provided a wide range of products for agricultural use. Technological advances are the most remarkable hallmarks of American agriculture, and the American farmers are among the most educated workers of the nation. Market demand, profit pursuit, technological advances and the well-endowed resource base together supported continued innovation and growing agricultural output. The United States became the world leader in agriculture production by the early 20th century.

Federal engagement in agricultural policies and Congressional legislation mushroomed during the 20th century.[12] With a check-and-balanced political system, the U. S. government's role in economic development is played out through all three branches: Congress makes laws, the executive branch implements and enforces the laws, and legal disputes are resolved in courts. The national consensus during rapid westward expansion was that the ultimate show of progress lay in unlimited and unmitigated economic growth. The West represents the greatest opportunities for individual and national endeavors.[13] Such a development model, especially when manifested in agriculture, has a mixed legacy.

The history of American agriculture, especially western agriculture, is characterized by a combination of forces aimed at commercial success. While the economic achievements have been remarkable and contributed to American national prosperity, there have been increasing concerns over sustainability and food security in light of problems with commercial agricultural practices based on an industrial model. Even more than its eastern counterparts, agriculture in the American West has been industrialized and export-dependent, as reflected in the larger scale of operations and the dominance of agribusinesses across the agricultural sector.[14] The American West has been at the forefront in addressing concerns over environmental impact of industrial agriculture and in pushing for more effective government conservation policy.

For China, the primary goals for Great West Development are enhancing national economic growth and strengthening security. These goals can only be achieved if the development is within the framework of long-term sustainability. Conservation must be the centerpiece of China's Great West Development, since ecological health has long been threatened in the Chinese West and resources are not nearly as "abundant" or "unclaimed" as in the old American West. In addition, there are complex cultural and political reasons for China not to pursue a "Wild West" policy. It must be noted that conservation is an alternative development model to the conventional growth-centered model; a conservation-centered development model can create economic opportunities, generate growth and improve living standards, at the same time maintain a relatively healthy and sustainable environment for long-term development. It may be more costly and yield less in the short-run than the conventional, growth-based model, especially when environmental costs are not figured in the total input-output matrix of GDP-defined economic indicators. Yet a conservation-centered model is consistent with the requirements for sustainable development, preservation of natural resources and environmental justice.

The evolution of agricultural conservation in the United States in general, in the American West in particular, illustrates both missteps and progress in the U.S. government policy regarding regional and agricultural development. Lessons from the U. S. experience should be examined by China closely as it is in the early stage of its massive western development. Should new agricultural development fail to maintain a healthy ecosystem in China's western regions, the prospect for China's Great West Development could be in jeopardy under the threat of environmental collapse, social instability and ethnic unrest.

ENVIRONMENT AND RESOURCE CONSEQUENCES OF NATIONAL POLICY FOR THE AMERICAN WEST

The rapid settlement of the U. S. frontier was directly attributed to agricultural success in the American West. In summing up the key reasons for the rise of the United States as the dominant economic power of the world, Wayne D. Rasmussen, a famous USDA historian, proclaimed: "The story of this great achievement is the history of American agriculture."[15] The American West was "won" not by the Winchester rifle, after all, but by the success of its agricultural economy powered by national policies such as the Homestead Act [16] and American industrialization. Dubbed "the first American Agricultural Revolution," the national policy that promoted large-scale, commercial agricultural settlement of the American West is considered a strategic move that changed the history of the United States and the world.[17]

The ethos of industrialization and *laissez faire* capitalism captured the essence of agricultural development in the American West. U. S. agriculture became the most productive and profitable in the world with rigorous innovation and application of the latest industrial technology to a well endowed resource base. By 1960, for example, one U. S. farmer could support twenty-two people with foodstuffs. The American West exemplified the industrialized agriculture. At the same time, however, profit-maximization pressure began to take atoll socioeconomically and environmentally. Family farms were squeezed out of business in large numbers. The eastern farmers and ranchers were not able to compete with the large-scale western farmers and were forced to the margins of the agricultural sector. The pursuit for profit maximization and industrial efficiency led to exploitative use of the land and over-dependency on exports. The former led to agricultural resource depletion, environmental contamination and product safety problems, while the latter resulted in detachment between communities and their food sources. All this happened under the watch of the federal government, whose policies were directly responsible for problems and crises that began to appear by the early 20th century. Among the worst environmental disasters was, as a result of development without conservation, the Dust Bowl.

The so-called Dust Bowl[18] was a decade-long rapid desertification in the 1930s, mainly affecting the Great Plains region. It resulted in devastation of cultivated land, uprooted many rural communities and prolonged the Great Depression in the region. It was due to the coupling of years of severe droughts and deep plowing that resulted in extensive soil erosion. Frequent dust storms, dubbed "black blizzards," overwhelmed many farms and decimated crops for a number of years, forcing foreclosures and displacement of many in the farming communities. The U.S. Department of Agriculture reported that, by 1934, roughly 35 million acres of formerly cultivated land had been destroyed; 100 million acres had lost all or most of their topsoil; and a further 125 million acres were rapidly losing their topsoil. By the late 1930s, two and a half million people from the Plains states had been forced to leave their home communities, most of whom moved west. The experience of these environmental refugees was captured by John Steinbeck (*The Grapes of Wrath*) and Woody Guthrie (*Dust Bowl Ballads*). The government responded to the situation with a series of short-term and long-term relief, including food delivery and cattle buying programs, soil conservation and afforestation. While the federal government responses were effective, they came after the Dust Bowl had already caused one of the worse ecological disasters in U. S. history, and its environmental impact has been felt ever since.

China must be extremely vigilant about the ecological impact of rapid development in its western regions, especially the arid areas that are still sparsely populated and may be tapped for settlement and development. China is already being confronted with serious environmental and resource challenges, and its western regions are at the forefront of such challenges. It must strike a delicate balance between the short-term economic gains and long-term environmental sustainability. National policy must clarify ecological limits and emphasize conservation, and must take into consideration environmental variations in the Chinese West. While a consistent theme in government policy statements has been conservation in Great West Development,[19] the focus of conservation has been on large-scale public projects such as land retirement, soil erosion control and afforestation. China must guard against blindly promoting industrial technology and operation in agriculture, which have caused American family farmers to lose control over their production decision making and have turned food production into an environmentally and socially exploitative business. The farmers became detached from their land and enslaved by the pressure treadmill of economies of scale. Should a similar process take place in China, the end result would be a defeat of China's efforts to conserve agricultural resources, encourage rural community development and relieve pressures on rampant urbanization. The Chinese West, for environmental and socioeconomic reasons, would be most devastated by this process. Again, the U.S. experience can be useful for China not only because of the environmental costs incurred by industrial agriculture but also the progress made in agricultural conservation in the United States in general and the American West in particular.

AGRICULTURAL CONSERVATION: LESSONS FOR CHINA'S GREAT WEST DEVELOPMENT

Conservation has become integral to sustainable agricultural development. China cannot afford to repeat the mistakes that the United States has made; rather, it can learn much from the U. S. experience in formulating and promoting agricultural conservation policies. Furthermore, China should try to set new standards for conservation and sustainable development in its Great West Development.

American agriculture has exemplified the triumphs and tribulations of private enterprise. The traditional model of U. S. agriculture was no different from any other business models based on competition through innovations and economies of scale, and with the threat of monopoly or oligopoly. Such a model has been extremely successful as measured by productivity and efficiency. The success of U. S. agriculture in the short-run, however, sowed the seeds for long-term problems. Among these problems is a wide range of environmental issues stemming from past and present practices. The consequences of these issues are so severe that many, not just the ones from the "left," have become concerned. The relatively fragile ecology of the American West bore the brunt of resource depletion and environment deterioration caused by industrialized agriculture. Extensive livestock grazing, which is practiced on nearly 70 percent of land, incurs enormous ecological costs,[20] and irrigation agriculture utilizes 90 percent of the fresh water developed for economic use in the American West.[21]

Agricultural development led the way for rapid westward expansion by the United States. Profit-pursuit and industrialization, as well as ignorance of the far-reaching impacts of unsustainable practices, caused severe resource depletion and environmental degradation. The American public became aware of the needs for conservation in the late 1800s, when industrialization brought out conflicts between economic growth and environmental deterioration. It is noteworthy that the concept and practice of conservation first gained momentum in the American West.[22] Yellowstone National Park (1872) in the state of Wyoming was the advent of conservation as a national policy, but the agriculture sector lagged behind in the early days of long-term conservation efforts. Conservation, especially in agriculture, first emerged as a grassroots movement and still depends on voluntary participation by farmers and ranchers for implementation and success.

The federal government was slow to take the lead, however, even to respond the demand from farmers and ranchers. Since then, an incentive-driven, private ownership-based conservation model has been making inroads in the United States. The U. S. federal conservation policies can be characterized as the promotion of connecting broad conservation interests with individual benefits and employing market incentives in setting agendas and achieving goals.[23] Specifically, these policies have forged a close relationship between conservation and price support.[24] The conservation policies have almost always been dependent upon incentive-based approaches for achieving environmental goals, rather than compulsion or penalty approaches (e.g., regulation and taxation). The overall trend of government conservation policy, at the same time, has been decentralization of policy formulation and implementation to local and state levels.

There has been a major shift in U. S. agricultural policy since 1996 as reflected in the 1996, 2002 and 2008 U. S. farm bills.[25] Before the 1996 farm bill, all farm bills used various schemes to pay farmers higher prices for a few selected commodities in return for reducing production by planting fewer acres. This was done to increase prices by control of production.[26] Since then, major U.S. legislation began to focus more on long-term conservation aimed at not only strengthening national and global food security, but also resolving severe environmental impact due to industrialized, commodity based agriculture.

Voluntary, market-oriented participation was made possible through promoting public awareness that links individual interests with societal benefits in conservation efforts. In addition, the perennial problems of overproduction and increase in capital input have encouraged agriculturalists to participate in conservation endeavors. By the 1960s, with Carson's *Silent Spring* (1962),[27] growing public awareness about environmental issues began to have an impact on agricultural policy. This was reflected in the 1985 Agriculture Act that contained a major expansion in conservation programs. Wetland conservation provisions (the so-called "swampbuster" provisions) prohibits farmers from draining wetlands in order to plant crops; those who do not comply would lose most of the benefits available to them under USDA programs. The key to the U.S. agricultural conservation policy, however, continues to be the market component and reverence for private property ownership, and institutionalized financial assistance to private agricultural operators.

Is the U. S. model applicable to China? The answer would have been a resounding NO not too long ago. Today, however, it has become clear that China's market economic reform is expanding and has become irreversible, and the Chinese government is determined to push for further reform, including privatization of agricultural land. Private enterprise appears to become the foundation for China's agricultural operations. The U. S. model of agricultural conservation, therefore, may become much more relevant to China. While the recent third plenary meeting of the 17th Chinese Communist Party (CCP) Congress (October 2008) did not reach a pivotal decision regarding land privatization, as had been anticipated, the Chinese government is set to implement much more liberal policies for land usufruct, including a seventy-year contract with nearly all the rights associated with actual land ownership.[28] There is evidence that land usufruct and other forms of quasi-property ownership in China have encouraged voluntary conservation efforts because of economic incentives to the farmers.[29] China may also need to further integrate pecuniary factors into its conservation policies and management approaches, and respond to grassroots demand for coordinating and rewarding private conservation endeavors.

The second useful lesson of the U. S. experience for China's western and agricultural development is the superior infrastructure for agriculture and rural development. Government services and resources are highly accessible to farmers and ranchers, such as agricultural subsidy programs and conservation methods such as land fallowing and output reduction. There is a research network of agricultural science and technology headed by the Land Grant colleges and universities. It has kept U. S. agriculture on the leading edge of technological innovations. Effective technology diffusion is then carried out by USDA's extension service networks,[30] 4-H youth programs,[31] and numerous channels for agricultural exchange and promotion such as county fairs. The superior infrastructure has been instrumental in the adoption of conservation-based agricultural practices such as no-till technique and crop-rotation systems. It may be the most transferable experience to China. First, it does not require significant political or cultural adjustments by China. Furthermore, China can tap into an already existing network of agricultural science and technology stations (*nongye keji tuiguang zhan*). Although under-funded and often poorly maintained, it is a rough counterpart to U. S. extension services. The Chinese government, however, may need to drastically increase funding in updating and strengthening this existing network, so that it would become an effective channel of technology diffusion for the farmers. In addition, China has a nationwide hierarchy of agricultural and forestry colleges. The Chinese government should improve and employ these structural and institutional resources already in place to promote conservation technology and encourage related investment in its Great West Development.

There are several lessons from American agricultural development that China can learn. First, China must try to resist the temptations of industrialized agriculture so that it could avoid the mistakes made in the evolution of American agricultural conservation. Economic growth during industrialization was a major driving engine for agricultural expansion in the United States. Industrialization drastically improved living standards in the rapidly growing industrial cities, which in turn provided a key market for agriculture and absorbed surplus labor in rural areas. It also spurred the development of food industry that integrated storage, processing, transportation and retail activities. While less than 2 percent of U. S. labor force is classified as agricultural today, the food industry is actually a top employer in the U. S. national economy. The flip side during the time was, in meeting or cashing in on new market demand, American agriculture became industrialized and shifted its focus to supplying a national diet that emphasized animal protein. This was especially the case in the American West where agricultural properties and operations were larger in scale and had access to the booming regional urban markets (such as those in the Midwest). Industrialized agriculture, as have been discussed, also resulted in negative environmental consequences due to its exploitative use of the resource base, which were particularly severe in the American West.

Rapid economic growth has presented great opportunities for the agriculture sector. These opportunities should not be captured only by traditional, profit-oriented practice; rather, they should also be tapped to stimulate development of new agricultural models that would take into consideration dietary and environmental health. There has been a shift of national diet toward much greater animal protein consumption in China. Such a trend would force agriculture to increase grain production in order to provide adequate animal feed, which would most likely be achieved through increases in cultivated cropland and capital input. In light of the drawbacks of industrialized agriculture that the United States has experienced, it is advisable that the Chinese government endeavor to steer the modernization of Chinese agriculture toward environmentally and nutritionally healthier structure and practices.

The second lesson from the American agricultural development is the characteristics of a representative democracy. There has always been strong political support for agriculture and rural development in the United States in the form of political lobbies, cooperatives and activist/grassroots organizations. All of them exist to protect agriculture and rural interests. The representative political system in the U. S. guaranties that its legislative body, the U. S. Congress, is responsive to voter sentiments and makes conservation legislation tailored to local conditions. At the same time, the U. S. political system can cause fragmentation, even inconsistency, in agricultural conservation policies because of the pressure of catering to local and regional special interests. The process of forming and implementing national policy through legislation is typically slow and sometimes inefficient. Nonetheless, the seemingly messy democratic political process can prevent major mistakes at the level of national government. In contrast, China has a highly centralized, non-democratic political power structure. Problems with this type of governance include lack of careful deliberation over national policies before their implementation and failure of taking into consideration regional and local conditions. In order to counterbalance the deficiency of the unitary political structure, the Chinese government should employ all available instruments to reduce the risks of rushing through far-reaching national policies, such as those for Great West Development. Academic research and debates, pilot programs, and government field investigations are some possible, albeit inadequate, substitutes to the democratic system's check and balance mechanisms.

Last, China may not be able to adopt the subsidy approach in agricultural conservation. One great contrast between the United States and China may be in the national demand on the agricultural sector, which presents quite different dynamics behind agricultural conservation policies. American agriculture has long been burdened by its "over-capacity," and oversupply has suppressed prices, shrinking the profit margin. Legislations and policies aimed at production reduction may have laid the foundation for modern agricultural conservation.[32] In China, oversupply may occur, if at all, in a specific product or season (e.g., cotton overproduction in recent years). Increasing agricultural output has been at the core of national policy, since China's agricultural supply has barely managed to keep up with population and economic growth. If the United States managed to coerce farmers to withdraw land from active cultivation as a *de facto* conservation measure, China may not be able to adopt this strategy at all. Yield and productivity increases may provide some room for land retirement, but increased yield and productivity are often associated with increase in capital, technological and chemical input. It is imperative that China not pursue industrialized agriculture, which has been proven to be unsustainable and which China cannot afford due to its negative impact on the ecosystem and on rural to urban migration.

The Chinese government's latest initiative for agricultural and rural reform is geared toward increasing economies of scale as the means of modernizing Chinese agriculture and empowering China's rural population.[33] The new initiative pushes for industrialization of China's agricultural sector, which can, as happened in the United States, transform an agrarian society into an urban-industrial society. In China's case, it can transform China's enormous agrarian sector with a population of nine hundred million. The inevitable environmental and demographic consequences of this initiative, however, conflict with China's imperatives of protecting the fragile environment and slowing down the massive rural to urban migration. Furthermore, agricultural industrialization in the United States has resulted in ecological deterioration at a scale that China cannot allow, especially in its ecologically fragile western regions. It has become evident that wholesale industrialization, dubbed modernization by the Chinese government, of agriculture will incur socioeconomic and environmental costs that China cannot afford. Is it possible to develop alternative agriculture that would address both economic demand and environmental concerns that China, especially its western regions, faces? In other words, is it possible for China to develop agricultural systems that strike a balance between economic growth and ecological limits, particularly in the Chinese West?

ALTERNATIVE AGRICULTURE: IS IT POSSIBLE FOR THE CHINESE WEST?

Sustainable agriculture and industrial agriculture are two fundamentally different philosophies, diametrically and irreconcilably opposed.[34] The industrialized commercial agriculture, while producing enormous quantity and highest yields of food, has caused severe damage to the environment and resources. It also threatens global food security and community health. Capital input and economies of scale continue to spiral out of control and farmers are driven off land in industrial agriculture.[35] China can ill-afford such wholesale industrialization of agriculture. What kind of model, then, should China adopt in order to balance short-term gains and long-term sustainability in agriculture? How can China keep farmers to the land with economic opportunities and living standards at least at the level that they, as migrant workers, could hope to have in the cities? The best marriage between economic growth and resource conservation is the development of alternatives to the dominant industrial agriculture. From a conservation point of view, alternative agricultural systems must be able to reverse the skyrocketing environmental costs of commodity oriented agriculture by promoting diversification of agricultural land use and outcome. In this sense, the emergence of alternative agriculture in the United States may provide the most useful inspirations to China's agricultural conservation and Great West Development, which is well-equipped to develop its own model of sustainable agriculture through revitalization of its tradition tailored to a rapidly industrializing and urbanizing society.

There is a resurgence of community-based agriculture in the United States stemmed from the Community Food Security movement.[36] Community Supported Agriculture (CSA)[37] is a key component of community based agriculture. Conservation and organic agriculture are at the core of CSA operations, which shun industrial management, processing and marketing. In the United States, the first known CSA was found in New England but the West Coast, especially California, has become a major concentration of CSA operations. There are two models for CSA arrangements, the shareholder CSA and subscription CSA.[38] The former operates more like a cooperative with members actively engaged in decision-making; the latter consists of a group of consumers that provide a farmer with money at the beginning of a growing season to provide the capital for growing costs. In both models, participants receive "shares" of the farm's production throughout the season. Shares often consist of boxes of goods produced by the farm, or CSA partners, delivered to neighborhood sites where members can pick them up.[39] There are about one thousand farming operations that meet the USDA's definition for CSA,[40] but over 12,000 farms marketed their products through CSA in 2007.[41]

CSA operations play an increasingly important role in promoting localized consumption, which may stimulate local economy, preserve local farms, and improve community relations. In addition, CSA members are said to enjoy a community that connects them to the production of their own food. Proponents argue that CSA farms are necessarily more sensitive to the needs and desires of the local community as they are more directly dependent on them.[42] Beyond this, CSA operations are exalted for their ability to create community and provide spaces for interaction based on moral convictions. In this way, the CSA experience comes to encompass much more than a simple market interaction and supports a wide range of values, ideals and practical necessities. There are different views, however, on CSA's ability of creating and maintaining food-based communities.[43]

The broad range of values incorporated into CSA has proved crucial to its continued economic existence in the capitalist economy where the key for profit maximization is to achieve economies of scale. Contemporary economic structures favor large economies of scale for efficiency in all aspects of product development and delivery in order to offer cheaper prices to consumers. Such structures have consistently displaced smaller competitors that were unable to capture similar efficiency and the subsequent price reductions. The persistence and growth of CSA operations, however, appear to present an alternative to profit-driven models for agriculture.

The emergence of CSA in the United States is directly linked to environmentalist movement and increasing awareness of dietary health. As a result, new agricultural policies have begun to take into consideration environmental health, food quality and security, and sustainable urban-rural communities.[44] These new policy imperatives are intended to remedy the ills of large-scale corporate agriculture. New and increasing market demands for organic food, coupled with government support, encourage more localized, labor-intensive agriculture systems, such as various forms of CSA, as alternatives to the industrialized agriculture. Alternative agriculture in the United States faces a major price disadvantage in competing with agribusinesses. Empirical evidence shows, however, that CSA operations can eke out a living serving a niche market in major metropolitan areas, where there is a large, affluent population that is attracted by the broad range of values CSA activities and products seem to encompass, especially its local community orientation and emphasis on resource conservation.

China may be well suited to resist wholesale industrialization of its agricultural sector and to foster alternative agricultural systems that are community-based and organic. First, the traditional Chinese food system was deeply vested in local community and *de facto* organic agriculture. Recent industrialization and modernization have not fundamentally changed the agrarian side of the Chinese dual economy despite some commercialization. Unlike the United States, special business interests are still relatively weak in supporting industrialized agriculture. Second, the fledgling but growing urban middle class has started to pay attention to their food safety and broad environmental quality and is increasingly demanding organic foods. While the Chinese West lags behind eastern, coastal regions in terms of urban and economic development, there are unique opportunities for alternative, locally-based agriculture. There are both demand and purchasing power for fresh and healthy foods at where there are major public projects are underway in the Chinese West. These projects receive large infusions of government funding, and the large workforce for these projects is usually paid very well. Both directly translate into local demand and spending.

As an alternative to industrialized agriculture, the revitalization of community-based agriculture in China can contribute significantly to the country's urgent need for slowing down massive rural to urban migration. Alternative agriculture such as CSA is labor-intensive. With both middle class demand and government price subsidies, now is the time for China to support alternative agriculture that is rooted in tradition and connected to modern economy. The ongoing global economic crisis has forced many migrant workers in the cities back to the countryside. The Chinese government could seize the opportunity and help facilitate agricultural systems that are community-based, labor-intensive, and ecologically sound. These systems would provide employment opportunities for many who otherwise may leave for the city again. The Chinese West may be the most promising testing ground for China's effort to reduce the pressure of urbanization in part because it is the western provinces that have generated the largest rural to urban migrations.

Furthermore, community based, organic agriculture is best qualified to serve the needs of environmental conservation and community food security in the Chinese West. While CSA and other alternative agricultural systems in the United States are largely private endeavors, the Chinese government can have a direct role in promoting alternative agricultural systems in counterbalancing the wholesale industrial model of agriculture. The central government should work with provincial and local governments to encourage and provide assistance to community-based agricultural activities that are already in place, such as gardening near cities and towns. By tapping even a small portion of its mammoth economic stimulus funds, the Chinese government can provide substantial support to local organic farmers so that they become economically viable, even competitive, in the expanding food market of the Chinese West.

China has no choice but to adopt conservation-oriented agricultural policies and promote alternative agriculture in its Great West Development. U.S. agriculture has surely achieved unprecedented success in output and productivity powered by the infusion of industrial technology, publically funded agricultural R&D activities, efficient channels of diffusion of innovations, and marketing networks. At the same time, the industrial model of agriculture has resulted in negative environmental impact and deteriorating community food security. In response, there has been a major shift in U.S. public policy that emphasizes conservation and alternative agriculture. The Chinese policymakers may need to closely the American agricultural experience, especially that of the American West. China should be able to formulate truly sustainable agricultural policies for its Great West Development by avoiding the mistakes that the United States has made and is now trying to correct.

NOTES

1. The official designation of China's western region includes twelve provinces and autonomous regions: Xizang (Tibet), Xinjiang, Qinghai, Inner Mongolia, Gansu, Ningxia, Shaanxi, Sichuan, Chongqing (municipality), Yunnan, Guizhou and Guangxi. See http://www.chinawest.gov.cn/web/index.asp (accessed July 27, 2009).
2. Announcement by the State Council on the Policies and Measures in Implementing Great West Development [Guowuyuan guanyu shishi xibudakaifa ruogan zhengce cuoshi de tongzhi], State Council Issue (2000) No. 33.
3. The geographic definition of the American West has evolved through time and there is still no consensus as to the eastern boundaries of the American West. See, Walter T. K. Nugent and Martin Ridge, *eds.*, *The American West: The Reader*, Indiana University Press, 1999. In this chapter, the American West broadly refers to the geographic area to the west of the Mississippi River in the continental United States.

4. "Reform and Opening Bring about a New West: Achievements and Prospect of Great West Development Strategy" [Gaige kaifang dailai yige xin xibu: xibudakaifa zhanlue chengjiu yu zhanwang], at http://www.gov.cn/jrzg/2008-10/13/content_1119760.htm (accessed on February 23, 2009).
5. "Growing a Nation: The Story of American Agriculture," *An Interactive Multimedia Program*, LetterPress Software, Inc., 2006, at http://www.agclassroom.org/gan/index.htm (accessed on November 19, 2008).
6. "An Act to Secure Homesteads for Actual Settlers on the Public Domain," 37th Congress (Session II 1862), http://www.loc.gov/rr/program/bib/ourdocs/Homestead.html (accessed on February 15, 2009).
7. "An Act Donating Public Lands to the Several States and Territories Which may Provide Colleges for the Benefit of Agriculture and the Mechanic Arts," 37th Congress (Session II 1862), http://www.loc.gov/rr/program/bib/ourdocs/Morrill.html (accessed on February 15, 2009).
8. There were several legislation in providing land grant status to colleges included the Hatch Act of 1887, the Smith-Lever Act of 1914, and the Elementary and Secondary Education Reauthorization Act of 1994.
9. "An Act to Aid in the Construction of a Railroad and Telegraph Line from the Missouri River to the Pacific Ocean, and to Secure to the Government the Use of the Same for Postal, Military, and Other Purposes," 37th Congress (Session II 1862), http://www.loc.gov/rr/program/bib/ourdocs/PacificRail.html (accessed on February 15, 2009).
10. Frederick Jackson Turner, *The Frontier in American History, With a Foreword by Ray Allen Billington*, New York: Holt, Rinehart and Winston, 1962.
11. The four projects are west-east electricity transmission, south-north water diversion, west-east natural gas transportation, and the Tibetan railroad.
12. Federal legislation appeared to have had several emphases: in the early 1900s, Congressional legislation focused on regulating agricultural products; by the time of Great Depression, emphasis of federal legislation shifted to supporting agriculture and rural communities. With post-World War II prosperity and as a part of the Great Society programs, agricultural economy became tied with social welfare. Conservation was first introduced by the short-lived Agricultural Act of 1956, but did not become the centerpiece of legislation until the 2002 Farm Bill.
13. For example, the New York Times had a provocative article titled "Farms in the Great West: Opportunities for Settlers up on the Public Lands," June 4, 1880.
14. Using the Mississippi River as boundary, the average farm size for eastern states is significantly smaller than that of the western states. The largest farms in eastern U. S. are found in Illinois, with an average of 352 acres in 2008. In contrast, the smallest average farm size in the western states was that California at 312 acres in 2008. Most, though, were larger than 400 acres and quite a few were more than 1,000 acres. See http://www.usda.gov/nass/PUBS/TODAYRPT/fnlo0209.txt (accessed July 11, 2009).
15. Wayne D. Rasmussen, ed., *Readings in the History of American Agriculture*, Urbana: University of Illinois Press, 1960.
16. Douglas W. Allen. "Homesteading and Property Rights; or 'How the West was Really Won,'" *Journal of Law & Economics*, 34 (1), 1991, pp. 1-23.
17. Keith Miller, "The West: Farming in the Great West (Part 2): Thresher's Coming, Putting Hay in the Loft," *History News Network*, March 27, 2002, at http://hnn.us/articles/656.html3-27-02 (accessed on February 18, 2009).
18. *The Dust Bowl*, at http://eh.net/encyclopedia/article/Cunfer.DustBowl (accessed on February 18, 2009). For a more complete study of the Dust Bowl, see Paul Bonnifield, *The Dust Bowl: Men, Dirt, and Depression*, Albuquerque, New Mexico: University of New Mexico Press, 1978.

19. A series of policy directives by the State Council repeatedly emphasized conservation in Great West Development. See, Announcement by the State Council on the Policies and Measures in Implementing Great West Development [Guowuyuan guanyu shishi xibudakaifa ruogan zhengce cuoshi de tongzhi], State Council Issue (2000) No. 33; Announcement by the General Office of the State Council on Suggestions by the Office of West Development Regarding Implementing the Policies and Measures of Great West Development [Guowuyuan bangongting zhuanfa guowuyuan xibu kaifaban guanyu xibudakaifa ruogan zhengce cuoshi shishi de tongzhi], State Council General Office Issue (2001) No. 73; Suggestions by the State Council on Further Promoting Great West Development [Guowuyuan guanyu jinyibu tuijin xibudakaifa de ruogan yijian], State Council Issue (2004) No. 6.
20. Thomas L. Fleischener, "Ecological Costs of Livestock Grazing in Western North America," *Conservation Biology*, 8 (3), 1994, pp. 629-644.
21. "Agriculture and Water in the West: What does the Future Hold?" Remarks by Bruce I. Knight, Chief, USDA NRCS, at a meeting sponsored by the Western States Water Council and the Western Governors Association in Salt Lake City, UT, September 28, 2004, http://www.nrcs.usda.gov/NEWS/speeches04/agandwaterinwest.html (accessed on February 10, 2009).
22. See "Chronology of Selected Events in the Development of the American Conservation Movement, c. 1850-1920," *American Memory Collections*, the Library of Congress, at http://memory.loc.gov/ammem/amrvhtml/ (accessed on February 18, 2009).
23. David Blandford and Richard N. Boisvert, "Multifunctional Agriculture—A View from the United States," plenary paper at the 90th Seminar of the European Association of Agricultural Economists Seminar: *Multifunctional Agriculture, Policies and Markets: Understanding the Critical Linkage*, Rennes, France, October 28-29, 2004.
24. Marc Ribaudo, LeRoy Hansen, Daniel Hellerstein, and Catherine Greene, "The Use of Markets to Increase Private Investment in Environmental Stewardship," *Economic Research Report* No. (ERR-64), September 2008, USDA Economic Research Service. Also see Marc Ribaudo, "Creating Markets for Environmental Stewardship: Potential Benefits and Problems," *Amber Waves: The Economics of Food, Farming, Natural Resources, and Rural America*, Vol. 6 (September 2008), Issue 4.
25. A key component for federal government to encourage agricultural conservation is the Environmental Quality Incentives Program (EQIP). See, "Emphasis Shifts in U.S. Conservation Policy," *Amber Waves: The Economics of Food, Farming, Natural Resources, and Rural America*, Vol. 1 (November 2003), Issue 5, updated July 2006.
26. Dennis Keeney and Loni Kemp, "A New Agricultural Policy for the United States," paper presented in part at the Advanced Research Workshop, Krakow, Poland, November 6, 2002.
27. Rachel Carson, *Silent Spring*, Boston: Houghton Mifflin, 1962.
28. "Resolution on Key Issues about Carrying forward Rural Reform and Development by the Central Committee of the Chinese Communist Party" [Zhonggong zhongyang guanyu tuijin nongcun gaige fazhan ruogan zhongda wenti de jueding], October 12, 2008, at http://cpc.people.com.cn/GB/64093/64094/8194418.html (accessed July 7, 2009).
29. Guo Qian, "Mechanisms for Rural Sustainable Development: The Effects of the Garden-Plantation Farming System in Shaanbei, China," *Journal of Sustainable Agriculture*, 13 (4), 1999, pp. 23-37.
30. The Extension Service (ES) is now part of USDA's "Cooperative State Research, Education, and Extension Service (CSREES)." U. S. Congress created CSREES through the 1994 Department Reorganization Act by combining the USDA's Cooperative State Research Service (CSRS) and Extension Service (ES) into a single agency.
31. The 4-H Clubs began as a rural youth organization with the goal of promoting agriculturally related knowledge and skills at an early age. It now gets support from

USDA under the Smith-Lever Act of 1914 and coordinates cooperative efforts by youth, volunteer leaders, state land-grant universities, state and local governments, 4-H Foundations, as well as USDA's Cooperative State Research and Educational and Extension Services (CSREES).

32. The first such legislation was the Agricultural Adjustment Act of 1933, Pub. L. No. 73-10, 48 Stat. 31(Originally cited as ch. 25, 48 Stat. 31). Facing worsened resource depletion and surplus problems, a series of new legislation were passed in an attempt to implement conservation measures, including the Soil Conservation and Domestic Allotment Act of 1935 (16 U.S.C. 590(e); 590(e)(1); 590(g); 590(h); 590(p)(1); and 590(q)(3); 49 Stat. 163), which established the Soil Conservation Service (SCS) as part of the Department of Agriculture.

33. Chinese government policy repeatedly emphasizes achieving economies of scale as one of the hallmarks of agricultural modernization and rural progress. See "Resolution on Key Issues about Carrying forward Rural Reform and Development by the Central Committee of the Chinese Communist Party."

34. John Ikerd, "Sustainable Agriculture: A Positive Alternative to Industrial Agriculture," paper presented at the Heartland Roundup in Manhattan, KS (A conference sponsored by the Heartland Network and the Kansas Rural Center), December 7, 1996. Also, see "Industrial Agriculture: Features and Policy," *Unions of Concerned Scientists,* http://www.ucsusa.org/food_and_agriculture/science_and_impacts/impacts_industrial_agriculture/industrial-agriculture-features.html (accessed July 10, 2009), and Benjamin Cohen, "Industrial Agriculture v. Sustainable Agriculture," *The World's Fair,* at http://scienceblogs.com/worldsfair/2009/05/industrial_agriculture_v_susta.php (accessed on July 12, 2009).

35. See Keeney and Kemp, "A New Agricultural Policy for the United States."

36. Formed in 1994, Community Food Security Coalition's mission statement was to bring about a situation "in which all persons obtain a nutritionally adequate, culturally acceptable diet at all times through local non-emergency sources." It has since become a nationwide movement and global initiative. See Andy Fisher, "What is Community Food Security?" http://www.sustainable-city.org/articles/security.htm (accessed on July 20, 2009).

37. "Community Supported Agriculture consists of a community of individuals who pledge support to a farm operation so that the farmland becomes, either legally or spiritually, the community's farm, with the growers and consumers providing mutual support and sharing the risks and benefits of food production," *Community Supported Agriculture* (USDA Publication), at http://www.nal.usda.gov/afsic/pubs/csa/csa.shtml (accessed on February 12, 2009).

38. Katherine L. Adam, "Community Supported Agriculture," *ATTRA Publication # IP289*, published in 2006 at http://www.attra.org/attra-pub/csa.html (accessed July 13, 2009).

39. Daniel Imhoff, "Community-Supported Agriculture: Farming with a Face on It," in Jerry Mander and Edward Goldsmith, *eds.*, *The Case against the Global Economy: And for a Turn toward the Local*, San Francisco, CA: Sierra Club Books, 1996.

40. See, Katherine L. Adam, "Community Supported Agriculture."

41. Based on data from USDA Census of Agriculture, 2007, which some dispute underestimated the number of CSA because it included those that would not participate in CSA activities, such as hobby farms, cash crop growers and other such operations. http://www.agcensus.usda.gov/Publications/2007/Full_Report/Volume_1,_Chapter_2_US_State_Level/st99_2_044_044.pdf.

42. Jack Kloppenburg, John Hendrickson, and G. W. Stevenson, "Coming in to the Foodshed," *Agriculture and Human Values*, 13 (3), 1996, pp. 33-42.

43. With eight CSA case studies, Cone and Myhre discovered that there was very weak participation in farm work and decision-making by CSA members, who therefore failed to develop any meaningful relations with farming and rural life. CSA to them is merely a self-satisfying way to obtain retailed food. Their conclusions cast doubts on the sustainability of CSA operations, as a result, since about seventy percent of all CSAs are subscription-based. See, Cynthia Abbott Cone, and Andrea Myhre, "Community-Supported Agriculture: A Sustainable Alternative to Industrial Agriculture?" *Human Organization*, summer 2000, http://www.encyclopedia.com/doc/1P3-55373357.html (accessed July 12, 2009).

44. "The 2008 Farm Bill Side-By-Side Comparison," *USDA Economic Research Services*, December 15, 2008, at http://www.ers.usda.gov/farmbill/2008/ (accessed on January 10, 2009).

Chapter 2

Grassroots Environmental Management in the Western United States: Can It Be a Model for Success for Western China?

Barbara A. Holzman and Qian Guo

This chapter examines grassroots environmental management in western United States through three case studies and its applicability to China's environmental management efforts. Grassroots environmental management has become a systematic approach in natural resources management and environmental protection. This approach focuses on the local level and aims for citizen participation. As a result, interactions among stakeholders in environmental management have largely shifted from confrontation to cooperation, which has led to informed environmental policies and better regulatory compliance. China and the United States face similar environmental challenges in their western regions despite their political and social differences. Grassroots environmental management may present a model for China to formulate strategies of environmental management in its Great West Development.

GRASSROOTS ECOSYSTEM MANAGEMENT: THE AMERICAN EXPERIENCE

China is embarking on an ambitious development of its vast western regions. Dubbed "Great West Development," the campaign is confronted by the fragile ecological conditions and rich, government-owned natural resources of the Chinese West.[1] China is applying a government-controlled, top-down approach in resource conservation and environmental protection that does not appear to work well.[2] What new approaches should China adopt in accordance with its continued market economic transition and inevitable political democratization? China may benefit from taking a close look at the rise of grassroots environmental management movement (GEM) in the United States, especially the American West.[3] In formulating new strategies for environmental management, China should closely examine how its current top-down approach compares and contrasts with the bottom-up approach embedded in the grassroots environmental management. In doing so, China may be able to avoid some of the missteps that happened to the environmental conservation movement in the United States and move forward toward a more community-based, more sustainable strategy for environmental management.

Natural resource management in the United States has evolved from disposal of public resources to private interests to protection of public resources for long-term, broad public interests. Early European settlers in the United States perceived the natural environment as object to claim and utilize. The dominance of private property rights, a cornerstone of European culture, was reflected in natural resource legislation. European settlers began to claim public land as early as 1785, when the Public Land Survey (PLS) was created to survey public land and divide it into sections, townships, and ranges as the basis for legal property ownership. As the United States continued its territorial expansion during the 19th century, a prevalent view emerged among American public that natural resources, especially land, were abundant. As a result, use of natural resources was exploitative with little thought about future needs and long-term sustainability of growth. Such a view was not only due to ignorance of nature's limit, but also one of the defining characteristics of capitalist value system. As environmental historian Daniel Worster pointed out, capitalism has been the decisive factor in the way that nature was "tapped" for economic gains in the United States. Capitalism refuses to recognize environmental costs as a component of real price of any product in order to maximize profit gains. Although many European settlers had communal heritage, the U.S. economic system coerced them to make individual wealth rather than communal survival as the highest priority.[4]

During its westward expansion, the U. S. government wanted to dispose of public land in order to settle the frontier as quickly as possible, both as a key strategy for geopolitical frontier control and for stimulating national economic growth.[5] The Homestead Act of 1862 granted 160 acres to U. S. citizens and legal residents who were willing to settle in the frontier.[6] The Homestead Act had the greatest impact on settlement to the west of the Mississippi River. The environment in the west was drier and less productive than that in the east, and the provision of homestead under the law increased in size to 320 acres in 1909 and 640 acres in 1916 to ensure western ranchers could operate with profit. Large tracts of land were given to railroad companies to encourage the building of a railroad system that could move products and people throughout the United States.[7] U. S. westward expansion created additional stresses to the land, which was less fertile and were often used as rangeland. A dramatic rise in cattle ranching displaced native grazers such as bison, antelope, and elk. Cattle ranching on public land had little oversight or management; overgrazing caused severe degradation of the public rangeland.

During the height of American industrialization, many began to realize the limit of the country's bounty, and something must be done if natural resources were to be sustained for future generations. The conservation movement was an important part of the Progressive Era (1890s – 1920s), when federal government began to be actively engaged in control of public lands for future economic use. In 1891, the Forest Reserve Act,[8] gave the U. S. President broad powers to set aside forest reserves on public land for conservation. It marked the beginning of the National Forests system. The national forests were to be managed in a sustainable manner so that they would continue to yield forest products in the long-run. Originally, all cattle must be taken off public land under the Forest Reserve Act, but this never happened. The particular measure of the Act was aimed at having the federal government manage public lands, but it was met with fierce opposition from ranchers and lumbermen in the West, who insisted that private enterprise on public land was their right. To them, the notion that the federal government would dictate the use of land in the West seemed unfair and was of little use to western states. This clash between national and regional interests in environmental management marked one of the first that would shape American conservation approaches. Western communities became entrenched in the belief that the states, not the federal government, should govern the resources within their borders.

The U. S. government continued to set aside public lands for both National Forests and National Parks.[9] Compared with other government attempts to reserve natural resources in public hand, the National Park system was established and expanded with less opposition from the private sector. John Muir led the efforts to set aside public lands as protected areas that must remain virtually untouched by human economic activities.[10] The so-called Preservation Movement was one of the first successful national campaigns by organized citizens to influence natural resource policies of the United States. The impetus for national parks came from a sector of the public that was quick to see that the valuable natural resources of the nation were quickly taken into private hands if the government did not specifically claim them for protection.

During the Progressive Era, federal policies regarding environmental management were formed based on federal government control. Little input from the land users or surrounding local communities was considered in the designation of national parks and national forests. At the same time, there was a lack of effective management on the public land by the federal government. As the United States continued to grow and prosper, pressure became to mount for tapping into public land for use by private enterprises. There was little accountability from extractive industries, which in turn maximized their activities on public land without costs. Increased grazing, mining, and forest harvesting in the American West left the large tracts of public land degraded and unusable. In response to the continued loss of valuable rangeland in the West due to overstocking and overgrazing, the U. S. Congress passed the Taylor Grazing Act of 1934.[11] The Act set up stricter management of public rangelands by limiting access and charging grazing fees. It included a community review component in the form of Grazing Districts and Boards that set local grazing limits. These districts provided opportunity for community input from one group of stakeholders, the ranchers. Limited opportunities for increased community participation were available in the 1930-1960s as the United States government continued to practice top down management of natural resources with the continued focus on extraction and production.

The Taylor Grazing Act was landmark legislation in environmental management policies of the United States. Not only was it effective in preserving the quality of public rangeland in the West, but also the first federal legislation that was proposed from the grassroots since the conceptualization of the legislation was by a group of Montana ranchers who had taken the imitative of improving the quality of both public and private rangeland. Also, the law calls upon the Secretary of the Interior to form local advisory board in determining grazing district boundaries and administering the law. In general, however, top-down, government mandated management of natural resources would continue, so did the tension and conflicts among the three groups, the conservationists, the preservationists and the advocates for state and private property rights. The antagonistic interactions among these major stakeholders became the key dynamic in the environmental history of the American West.

The 1960s brought in a decade of public empowerment through civil rights and anti-war movements and also influenced the environmental movement. Public awareness of environmental issues increased and demands for a better environment began to have an impact on government policies. The first law that addressed the need for public participation was the National Environmental Policy Act (NEPA) of 1969.[12] NEPA set out a process to incorporate citizen responses to proposed land management and development by the government as well as by private entities seeking permits from the government. The process initiated a new era for public land management, where the public was able to respond to government plans. While the public was now able to comment on proposed government actions, the government was largely irresponsive to public opinions. In 1973, the Endangered Species Act (ESA)[13] gave the public more power by providing citizens with the legal power to block federal action if endangered species included in the law were threatened. Environmentalists also moved into new battlegrounds of demanding for clean water, clean air, and safe and healthy public lands. The federal government responded with a series of protective laws pertaining to clean water, clean air, with regulations on development and hazardous materials. These laws designated proper use of the nation's resources in an effort to improve the current environment.

With the environmental movement making important inroads into public policy, management of public lands and their resources came under closer scrutiny as the public demanded better land management, more public access and public input. Land management began to shift to exploring multiple uses of public land, in contrast to the past, largely extractive uses such as cattle ranching. Public land management agencies, such as the Bureau of Land Management (BLM), were held accountable for protection of wildlife, watersheds, and wilderness. Federal resource management plans required an opportunity for public response as established by NEPA, as well as several other land use acts, such as the National Forest Management Act (NFMA)[14] of 1976 and the Federal Land Policy and Management Act (FLPMA)[15] in 1976. Land use and management strategies were still top down, however, as the government agencies made most of the decisions regarding public land use with little local and community input.

While public involvement began to increase in the early 1970s, its role remained peripheral in the top-down management apparatus. Communities were given a voice, but the government was not compelled to respond to the voice. From the 1970s to late 1980s, local communities continued to show eagerness in participating in the management of local resources and were very critical of the federal management practices. Many of the resource decisions were still coming from Washington D.C., often in response to demands of urban dwellers concerned with the protection of public lands, with little consideration of the needs of local rural communities using the land. As a result, community resentments toward ineffective and irresponsive bureaucracy in Washington, D. C., converged to form popular movements in the 1970s and 1980s.

Dubbed the Sagebrush Rebellion and Wise Use movement, these popular movements were motivated to counter the perceived federal ("Eastern") bureaucracy that had been directing land management increasingly influenced mainly by large environmental lobbying groups.[16] The issue of state rights vs. federal power resurfaced, reminiscent of the opposition to the Taylor Grazing Act decades earlier. The Sagebrush Rebellion demanded that the federal government give more control of western public land to state and local authorities to increase the growth of western economies. Rural land users challenged the current strategies of natural resource management and demanded that local users be part of the decision making processes that determined local land use and management. The more rural western states argued that decisions by agencies in Washington D.C. neglected the economic and cultural needs of the local community. Early interactions between the federal government and local communities were confrontational, with the use of lawsuits, protests, barricades, and harassment of government land managers.

Urban environmentalists saw these rural protests as an effort to take over public land for private uses and demanded direct involvement in the matter. Further conflict ensued, as a result, with the debate centered on local economic interests, mainly extractive use for local economic gain, versus broad public interests, i.e., protection and preservation of biodiversity and endangered species. Environmentalists saw local attempts as land grab efforts while local communities saw the environmentalists as outsiders who tried to limit their economic development. Sagebrush Rebellion held strong political support and was associated with private business interests. While the sagebrush rebels included citizens of rural communities, a key driving force was private corporate interest, which wanted to secure extraction rights and maximize profits without concern of the environmental health of federal land. In the U.S. Senate, an effort to move public lands into the hands of the states emerged. Utah Senator Orrin Hatch introduced a bill in the Senate (S. 1680) in 1979 that would transfer land from federal to state ownership, one of the main goals of the Sagebrush Rebellion. The bill failed to pass the Senate.

Known for his pro-business stance, President Ronald Reagan was seen to side with rural communities in the West, declaring that he too was "a sagebrush rebel." Upon taking office, Reagan quickly attempted to overturn environmental legislation that limited extractive uses on public lands. He halted efforts to designate many public land holdings as protected wilderness areas where many economic uses would be prohibited. Reagan's policies on land management outraged the environmental community. Alarmed by the inroad made by opposition, local groups in urban and rural communities began to form in advocating causes of conservation, protection of public lands and endangered species and demanding that government agencies consider the needs of the communities.

By the 1990s, it became clear that the traditional government approach of top-down, agency-led planning and decision-making had failed to effectively protect the fragile ecosystems and led to increasing tension among stakeholders and management gridlock. The Clinton Administration (1993-2001) ushered in a new inclusive perspective and set about creating avenues for collaboration among interests groups in hopes of limiting lawsuits and moving their agenda forward. The Environmental Protection Agency (EPA) initiated the "Common Sense Initiative (CSI).[17] CSI brought together industries and regulators to guide efforts on environmental protection through consensus-based decision-making. This approach was first implemented at the manufacturing sections, e.g., iron and steel industry and automobile manufacturing. This approach influenced natural resource management in the United States. Government agencies such as the forest service and park service were mandated to use an "ecosystems approach" to management, [18] which required a more holistic approach to managing the nation's resources. In ecosystem management, the ecological importance of the land is considered along the economic and cultural needs of the stakeholders. [19] In ecosystem management, all stakeholders or their representatives were part of the discussion and the goal was to come to consensus regarding how best to manage the land. Stakeholders include the federal government, local and state governments, local landowners and users, environmentalists, and additional parties or individuals that have some stake in the outcome of management.

Adoption of this new, collaboration-oriented approach was particularly challenging in the western states, where tension had been high between private landowners and property rights activists on one side, and government regulators and conservation organizations on the other. For years, many local communities in American West had depended on extractive activities such as mining, livestock grazing, and timber harvest for their livelihood. Unfortunately, these practices often conflicted with the goals of conservation. Once moving beyond old animosities, however, stakeholders in environmental management came together in an attempt to develop better land management practices that would address the economic needs of the community and the sustainability of the ecosystem. Eventually, the new approach to environmental management became more sophisticated and accepted, because of its design to address some of the challenges in earlier land management dilemmas. Dubbed grassroots environmental management, the new approach integrates a number of related concepts and practices, such as community based ecosystem management

(CBEM), grassroots ecosystem management,[20] and participatory environmental policy.[21] In grassroots environmental management, resource management stems directly from a coalition of local users, as well as local, state, and federal agencies. It is an attempt to bring as many parties to the table as possible so that the resultant management plan would be beneficial to all parties.[22] In CBEM, the community is part of the management unit and management plans are based on consensus.[23] The intent is not preservation of resources or sustainable use, but a combination of ecological, economic, and cultural uses that encourage economic development that is abide by the needs of sustainable use of the land resources along with the need to protect certain resources in a larger ecological context.

One of the early attempts to manage public natural resources using this approach was by the Clinton Administration in the early 1990s. The proposal was to create a plan for the management of the national forests in the Pacific Northwest that would address stakeholders concerns for conservation, endangered species and watershed protection, as well as sustainable forestry and increased economic development.[24] The taskforce brought together local users, first nation representatives, conservation organizations, federal land management, environmental agencies, and timber companies. In the end, the Northwest Forest Plan proposed options for management. Unfortunately, the final form of the plan was not one of the alternatives reached by group consensus and resulted in lawsuits delaying plan implementation. While the success of the plan is debatable,[25] the process was seen as being effective since it created a structure in which coalitions could be formed and strive for consensus in land management decision making.

Similar attempts by the federal government enabled local groups to have their voices heard in local land use policies. Many similar public policy forums have emerged; some are coordinated by government agencies, while others were formed out of opposition to government plans that lacked input from local communities needs. These community–based collaborations have a variety of structures, such as citizen advisory boards, taskforces, and watershed councils; all have gained widespread popularity. The local community's connection to place was an important aspect of community based ecosystem management. Reliance on local natural resources mobilized people to share in its governance. Increased local participation was thought to lead to better management because community members have greater interest in the sustainability of their local resources, and may be better able to identify the multitude of current uses of the land.[26]

Collaborative, decentralized management, as represented by GEM, represents a new era of resource conservation in the United States. This is particularly true in American West, where the federal government owns a large percentage of the land, and where, like authors Harmon and McKinney observed, the most rigorous debates, clashes, experiments and collaborations over public land and resource management have unfolded.[27] Grassroots environmental management is a meld of many of the previous land management strategies, including those of preservation, protection, and conservation with those of economic development and sustainability. It is thus an effective approach to managing publically owned natural resources since it conforms to interest-based negotiations that have characterized the dynamics of environmental history of the American West.

Theoretically, with such community support and consensus, government agencies would find it to implement management plan. [28] In reality, implementation of environmental management policies are never simple or easy, since grassroots coalitions attempt to control and maximize their individual interests while maximizing collective goals within the constraints posed by local, state, and federal agencies. [29] There are thousands of coalitions in American West that function at different local or regional scales with varying extent of success. Many researchers have provided detailed case studies of these groups in assessing their effectiveness and the outcome of grassroots environmental management.[30] The following four case studies demonstrate how grassroots efforts operated and interfaced with other stakeholders in local ecosystem management. Each represents a sub-type of community-based resource management effort and all attempted to adopt approaches similar to the conceptual grassroots environmental management.

The Blackfoot River Challenge

The case of watershed management along Blackfoot River, Montana, is a classic study of how grassroots, cooperative approach could work well for all stakeholders in environmental management.[31] With 1.5 million acres, the idyllic landscape in the watershed of Blackfoot River is matched with rich biodiversity and other natural resources. At the same time, decades of unsustainable human activities, such as poor mining, grazing, and logging practices have resulted in water quality and water supply issues, sedimentation, and a declining fishery, invasive noxious weeds, degradation of private lands and water rights disputes. Increased large-scale cattle ranching, resorts, and other commercial development also posted threat to the rural characters of the valley. The residents of the Blackfoot River watershed adopted a grassroots approach and created partnerships among area residents who shared a common goal of preserving the watershed. In 1993, local citizens and representatives from state and federal natural resource agencies created the Blackfoot Challenge, a nonprofit organization dedicated to promoting cooperative conservation efforts in the Blackfoot Watershed. Rather than using ineffective regulatory threats of the past, the U. S. Fish and Wildlife Service empowered local landowners to improve resources on their lands and meet natural resource objectives, while maintaining their rural lifestyle activities such as ranching, hunting, fishing, and forestry.

There are over 660 partners in the Blackfoot River Watershed Advisory Group, including private landowners, environmental groups, local, state and federal agencies, timber companies, and private foundations.[32] Together, they have placed 90,000 acres of private lands in conservation easements, developed stream flow protection plans, and undertaken 500 projects with 147 landowners restoring 38 miles of streams, 62 miles of river and stream banks, 2,600 acres of wetlands, and 2,300 acres of grasslands.[33] The group also worked to reduce human-wildlife conflicts by introducing preventative and proactive strategies and projects.

The cooperative approach taken maintains the economic well being and sustainability of the community, while allowing for the conservation of biodiversity and the larger ecosystem. The coalition was successful in meeting the goals of the local community while abiding by federal land management laws. Federal agencies continue to play an important role in supporting the efforts of the watershed group. The case of Blackfoot River Watershed management has become a classic success story for grassroots, cooperative environmental management. It has won numerous awards and grants, was featured by Public Broadcasting System, and lauded by the federal government. As Blackfoot Challenge website proudly announces:

> Cooperation. Trust. Consensus: Three words guide the Blackfoot Challenge. Here, private landowners take the lead and public agencies follow in a shared goal-to keep large landscapes intact and rural lifestyles vital. What's our secret to success? Focus on the 80 percent that folks can agree on and not the 20 percent that might divide us. Leave egos at the door and wear a Blackfoot Challenge hat in our meetings. Listen well, don't rush and find the common ground before acting.[34]

The Henry's Fork Watershed Council

In Idaho, a successful grassroots effort was initiated with the formation of the Henry's Fork Watershed Council. The Henry's Fork watershed lies in eastern Idaho and western Wyoming. It includes 1.7 million acres and over 3,000 miles of rivers, streams, and canals.[35] The watershed supports healthy populations of fish and wildlife, including several threatened and endangered species, as well as high-quality recreational experiences.

About 40,000 people live in the area within the watershed, which includes Yellowstone and Grand Teton National Parks.[36] In the late 1800s, the original settlers built irrigation canals and storage reservoirs for water. The canals diverted water from several of the rivers and now irrigate over 200,000 acres of farmland. Other important parts of the economy include recreation and tourism in the nearby National Parks as well as energy, timber, and livestock production.

As interest in the area grew, competing demands for irrigation and hydropower, as well as instream flow needs for fisheries and recreation increased. At least 25 federal, state, and local agencies have management or regulatory jurisdiction in the Henry's Fork Basin, a situation that contributed to fragmented planning and decision making. According to the Partnership Resource Center of Henry's Fork Watershed Council, the lack of agency coordination was hindering progress in addressing soil erosion, water delivery, and water quality problems, thereby worsening rather than solving problems in the basin.

In 1993, Henry's Fork Watershed Council, a grassroots organization, was formed to craft a new, non-adversarial approach to reconciling watershed issues in the Henry's Fork Basin. The Council is comprised of citizens, scientists, and agency representatives, who reside, recreate, make a living, and/or have legal responsibilities in the basin. In their mission statement, the Council identifies itself as "a grassroots, community forum that uses a non-adversarial, consensus-based approach to problem solving and conflict resolution among citizens, scientists, and agencies with varied perspectives."[37] The Council addresses social, economic, and environmental concerns in the basin, by respectfully cooperating within the group and by abiding by federal, state, and local laws and regulations. The Council's goal was to restore and enhance watershed resources as well as maintain a sustainable watershed resource for future generations. The group worked with and within the current government structure and was recognized and chartered by the state of Idaho and continues to be the leader in the stewardship of the area. The Council also has the responsibility to identify and coordinate funding sources for research, planning and implementation and long-term monitoring programs. They serve as an educational resource to the Legislature and the public. This successful grassroots coalition continues to be the main decision-making body on land management issues within the watershed.

The Quincy Library Group

The case of Quincy Library Group (QLG) demonstrates how environmental advocacy groups could effectively influence federal legislative process by adopting a community-based, grassroots strategy. QLG was formed in 1992 by representatives from different interest groups that have stakes in federal forest management in Lassen, Plumas, and Sierra Counties in northeastern California.[38] The stakeholders came to recognize that the so-called "timber war" had been waged at the loss of all involved. QLG stakeholders intended to cooperate with rather than confront each other to find compromises when all needed to face new plans of forest management proposed by U. S. Forest Service (USFS). The conflict in the 1990s in the area was over a forest management plan for the national forests surrounding the local communities. The environmentalists were concerned about protecting old growth forests and an endangered species, while the timber industry who were supported by the local employees were looking to continue timber production and the economic well being of the community. Realizing the inevitable gridlock if the stakeholders had continued to clash with each other, QLG attempted to coordinate in opposition to the USFS plan, despite the fact that the stakeholders in the group might be opposed to different provisions of the plan. The group expanded to include members of the timber industry, municipal and county governments, fisheries and environmental groups, and local citizens.[39] As a sign of searching for consensus, the group's stated goal was to promote forest health, ecological integrity, adequate timber supply, and local economic stability, allowing local communities to survive economically, yet afford adequate environmental protection for the forest and endangered species living in the forest.[40]

After two years of sometimes tumultuous discussions, a pilot project for managing the forests was created and supported by the group. USFS agreed to only part of the plan which allowed for timber sales and economic growth in the community, along with set asides for ecological value (i.e. old growth, endangered species habitat, roadless areas), and the restoration of riparian areas and other disturbed habitat. Because the Forest Service would not agree to all aspects of the collaborative plan, the group convinced their local Congressman and their U.S. senator to bring their pilot project to Washington in the form of an Act. The Herger-Feinstein Quincy Library Group Forest Recovery Act was passed and signed into law by President Bill Clinton in October of 1998.[41] Senator Feinstein commented at the time that the legislation was a "real victory for local consensus decision making" and proved that very difficult environmental issues can be solved if people worked together toward a common goal.[42] The Act proposed the pilot project be implemented and overseen by the QLG and evaluated at five year intervals. It took several years of before the Act could be implemented due to litigation and the need for clarification of management implementation. The results and success of the pilot project are still pending.

The QLG is an example of a collaborative process where local management overrode that of the federal agency. If successful in promoting economic growth and protecting habitat, the results would be encouraging for other like groups. The power of the group to go above the U.S. Forests Service, the agency charged with management of the forests, was a new development in the collaborative process that was met with caution. Local groups allowed to manage public lands by an act of Congress, in opposition to the federal agency charged with their management, at the cost of taxpayers, and with little accountability for their actions is controversial.[43] Complaints surfaced about the timber industry heavily influencing the group, and provides insight into the possibility of a biased decision making process. The potential for these community groups to be taken over by large commercial interests is evident. Diligence on behalf of all interested parties and the government must be insured for equitable solutions.

The California Rangeland Conservation Coalition (CRCC)

The last case is a statewide effort to deal with land management issues that affected a targeted ecosystem and the cultural heritage of the state of California. The California Rangeland Conservation Coalition was formed in 2005 and is comprised of environmentalists, ranchers, university researchers, and resource professionals from federal and state agencies. Members of the group were to work together on issues of land use in the rural grasslands and woodlands that are presently used for ranching. In the past, cattle grazing were considered contrary to the goals of environmental protection of watersheds, endangered species, and ecosystems. This notion, however, has changed.

With the growing population in California and recent efforts to subdivide rural areas for residential housing, rangelands in the state are quickly disappearing. With market pressures and changing zoning laws, ranchers find their livelihood endangered, while conservationists recognized that losing these rangelands would drastically reduce open space and habitat for wildlife. Subdivisions or total loss of these ecosystems was seen as extremely detrimental to the biodiversity of California. Together the two groups, that were once adversaries, joined state and federal agencies to acknowledge the importance of rangelands in California and to work to protect these lands. The Coalition drafted a resolution documenting common ground for the conservation of the rangelands encircling the Central Valley, including the Sierra foothills and interior Coast Ranges of California.[44] The California Rangeland Resolution recognized that these rangelands and the diversity of species they support is largely due to grazing and other land stewardship practices of the ranchers that own and manage them. The importance of maintaining these lands was an important goal of the diverse groups within the coalition.

The Coalition continues to grow and includes over 75 agricultural groups, environmental interest groups, as well as state and federal agencies. The regional groups of the Coalition work on regional issues and the statewide coalition convenes a summit at least once a year to work together in achieving their goals by creating an action plan. The action plan provides the foundation for coalition members to acquire funding for conservation programs, coordinate permitting processes, garner support for cooperative conservation projects, fulfill research gaps, and provide landowner assurances and incentives for ecosystem services. Environmental groups and land use agencies are pleased with this coalition and acknowledge their common goals. The Coalition is becoming a growing voice in land management issues in California. The success of this grassroots effort to bring together once adversarial groups to work on a common goal provides further evidence that community based ecosystem management can be a successful model for in guiding future land management.

Summary

Although the cases presented above were for the most part successful, other attempts at community based management have been less so. Hibbard and Madsen reported that two of the three cases they studied were considered problematic.[45] They felt such groups could be biased by the overwhelming influences of commodity based interest such as the timber industry, which may have been the case in the Quincy Library Group.[46] They note that economic concerns can overshadow ecological concerns and therefore lead to inequitable decisions on how the land should be managed. Walker and Hurley report on a collaborative process that was thwarted before it even begun. They argue that local politics should be a major consideration as political groups may see these collaborative processes as a way to further their political agendas, with little regard to the management issues at hand.[47]

Yaffe and Wondolleck caution that these coalitions are often made of volunteers with limited time to work on the issues, and those who may be interested but unable to volunteer their time or attend meetings may be excluded. In addition, since these coalition meetings are often held within the local community, interest by others who are not local can be left out.[48] While NEPA and the Forest and Land Management Act that followed it ensured that the public could review the planning documents and provided a process for interested parties to comment, plans created by local coalitions do not always provide the same accessibility. In the case of the Quincy Library Group, by going to the legislature to get their plan made into law, they skirted around the objections of the Forest Service and the process of open public comment.

Resistance to grassroots ecosystem management planning is often based on the issues of conflict between local communities taking on the management of public resources, while the costs are incurred by the public, and ultimately the responsibility for proper management lies with land management agency. Loomis argues that local residents may lack the multidisciplinary and technical planning expertise to create a feasible and equitable management strategy. He notes that management by local groups is problematic because they are not held accountable for the plan, and it is the taxpayers who must bear the cost, but have little influence on how the land is managed.[49]

Successful grassroots collaborative management plans on the other hand are made when there is enough interest generated to allow for a large group of participants who feel that they have some direct decision making powers.[50] Successful coalitions and the management plans they create are based on the clarification and agreement of similar goals, as well as the willingness to compromise.[51] Stronger coalitions are created when relationships are formed among interested parties, and these parties willingly take responsibility and ownership of the process and product. Government supported initiatives with funding and research attached such as the Black River Watershed Challenge and Henry's Fork Council are more successful.

CONCLUSION AND LESSONS FROM THE UNITED STATES AND CHALLENGES TO CHINA

U. S. conservation has a history of grassroots action because of its political culture (distrust of government) and economic system (private property rights). Rapid economic development, in large part thriving upon rich natural resources in the United States, allows the country to prosper. Until recent decades, conservation policy and approaches were top-down and strongly oriented toward promoting agriculture, resource production and other economic activities.

Conservation in the American West appeared to have headed in two directions. While the preservationist movement helped create the National Park System and protect large tracts of public land from rampant commercial development and privatization, there was limited conservation effort at the grassroots level for private properties and other public lands, resulting in unsustainable uses and severe environmental damages.

By the mid-20th century, the environmental movement in the United States started a new impetus to conservation and a major paradigm shift began to take place. The federal government began to tilt toward much more decentralized management. Meanwhile, grassroots efforts were becoming a major force in conservation as public awareness of sustainable development increased markedly. Grassroots environmental management (GEM), also known as community based ecosystem management (CBEM), emerged as one of the most effective approaches in conserving public resources for sustainable and collective benefits. This approach, as the four case studies above demonstrate, brings together as many stakeholders as possible and strives for compromises between long-term environmental health and short-term economic viability, as well as between public and private social and economic interests. How to foster GEM may be the most difficult challenge of the environmental tasks in China's Great West Development; it may also reap the greatest long-term benefits for China if GEM becomes an integral part of development. Despite the challenges, there are several aspects that are reasons for optimism for GEM to succeed in China.

First, China fully realizes that it must strike a balance between economic growth and environmental health in order to achieve long-term sustainability in its Great West Development,[52] and it has always been a centerpiece of Chinese government development policy,[53] despite the fact that conservation policies have often been ambiguous and implementation weak. The unwavering determination of the central government in finding effective approaches of environmental management bodes well for models that are tested and successful. As there are certain parallels between the Chinese West and the American West, grassroots environmental management (GEM) in the American West should provide some transferable lessons. As China continues its economic transition to market economy, private property rights would become more and more embedded in the society, providing the foundation for institutionalized grassroots efforts across the spectrum of China, providing impetus for grassroots environmental management.

Second, China's highly centralized political structure makes China's environmental management a classic case study of top-down, heavy-handed government conservation approach. On the one hand, this political reality may make decentralization of power related to conservation more difficult. On the other hand, the Chinese government and academic community have concluded that the top-down approach has not worked well, and new approaches need to be developed. The U. S. experience in this regard is most appropriate reference to China, especially as local economic and even political autonomy is becoming a strong trend.[54] The question is whether and how the Chinese central government would loosen its iron grip on power, allowing local governments and communities to have much greater input in the decision-making in managing their immediate environment and resources. The trend, though, does point to greater liberalization and inevitable decentralization of power.

Third, China is becoming an increasingly open society, which means that it is possible for direct international involvement in fostering grassroots environmental management efforts with persistent and culturally sensitive international assistance. China's fledgling grassroots environmental organizations are grouped together with non-governmental organizations (NGOs).[55] They are formed and maintained by kindred souls, and mainly draw membership from urban areas (e.g., Friends of Nature, the first environmental NGO in China founded in 1994).[56] This situation almost mirrors the case in the United States from the late 1800s to early 1900s. The question is whether these organizations are "grassroots." If we regard true grassroots as being anchored upon local and individual participation, which is more likely to be sustainable, then many of the high-profile environmental organizations in China may not qualify. These organizations may not supplement the role of true grassroots, locale-specific environmental or ecological management. Rather, like in American West, they as "outsiders" often clashed with the local community and their efforts were often unsustainable.

Last, China may actually have a favorable condition for fostering GEM with its long-standing culture of *de facto* grassroots conservation, which was key to having sustained Chinese civilization for millennia.[57] On the one hand, Chinese population pressure frequently pitted the Chinese against nature, and the government resorted to outward expansion for living space. On the other hand, at the community (township and village) level, the Chinese were vigilant about managing natural resources and maintaining environmental health in order to sustain themselves despite of the fact that the government often lacked consistent policies and regulations. Terracing and fish pond-mulberry bank system are but two examples of traditional community-based agricultural conservation measures that worked well for hundreds of years in China.[58]

There is an emerging, albeit embryonic, grassroots environmentalist movement in China. A number of conservation organizations have the potential of being truly grassroots and effective in reaching their goals and the Chinese government has provided a formal registration process to these organizations. One of these is "the Snowland Great Rivers Environmental Protection Association,[59] which is aimed at encouraging the local people to know more about their environment and its development. It attempts to facilitate information about the traditional ecological understanding and approaches of peoples on the Tibet Plateau. This organization does have government participation but its membership consists of mostly local Tibetans who have an urgent interest of preserving the ecosystem near the origins of the Yellow and Yangtze Rivers for their own survival. It will be interesting to see how this grassroots organization evolves and forges a mutually inclusive relationship with the local government, and how effective it is in accomplishing its own goals.

It is to the interests and benefits of China in its Great West Development to promote grassroots environmental organizations as the steadfast safeguards for local, fragile ecosystems, as the case in Western U. S. has shown. The government and society in general must support and facilitate grassroots efforts in managing local ecosystems. Grassroots environmental management has several apparent benefits for any society, especially an emerging China: (1) participants are out of conviction and self-interests, their endeavors are a more sustainable form of conservation, and they can be more effective by circumventing bureaucracy, (2) it is a strong stimulus for building civil society and contributing to democratic transformation through popular participation, and (3) such a movement and its specific operations are usually most informed and familiar with and responsive to local changes and needs.

Grassroots movement of any kind must come from awareness and protection of self-interests. In order for China's grassroots environmental management to becoming institutionalized, several factors need to be in place. First, in order to have grassroots conservation movement, individual and cooperative property rights have to be guaranteed, as some case studies in Chinese West have documented.[60] Second, the benefits of having grassroots conservation must be shared by the local participants, and legal and administrative apparatus need to be developed to guarantee such sharing. China can and must move toward a more grassroots based environmental management structure,[61] and avoid the missteps that the United States made in its own western conservation endeavors. The task is challenging but China has no better alternative, especially in continuing its ambitious Great West Development. China, in this regard, can learn much from the lessons from the American West.

NOTES

1. In this chapter, the Chinese West follows the official designation of China's western region includes twelve provinces and autonomous regions: Xizang (Tibet), Xinjiang, Qinghai, Inner Mongolia, Gansu, Ningxia, Shaanxi, Sichuan, Chongqing (municipality), Yunnan, Guizhou and Guangxi, see, http://www.chinawest.gov.cn/web/index.asp (accessed July 27, 2009).
2. Stefanie Beyer, "Environmental Law and Policy in the People's Republic of China," *Chinese Journal of International Law*, No. 5, 2006, pp. 185-211.
3. The geographic definition of the American West has evolved through time and there is still no consensus as to the eastern boundaries of the American West. See, Walter T. K. Nugent and Martin Ridge, eds., *The American West: The Reader*, Indiana University Press, 1999. In this chapter, the American West broadly refers to the geographic area to the west of the Mississippi River in the continental United States.
4. Donald Worster, *Rivers of Empire: Water, Aridity, and the Growth of the American West,* New York: Pantheon, 1985.
5. Cary Coglianese and Laurie Allen, "Does Consensus Make Common Sense? An Analysis of the EPA's Common Sense Initiative," *Environment*, 46 (1), 2004, pp. 10-25.

6. 37th Congress (Session II 1862), "An Act to Secure Homesteads for Actual Settlers on the Public Domain," at http://www.loc.gov/rr/program/bib/ourdocs/Homestead.html (accessed on February 15, 2009).
7. The legislation that granted public to railroad companies was the so-call Pacific Railroad Act of 1862, see http://www.loc.gov/rr/program/bib/ourdocs/PacificRail.html (accessed on February 15, 2009).
8. Passed by the 51st Congress (Session II 1891), the full name of this act was "An Act to Repeal Timber-Culture Laws, and for Other Purposes" (26 Stat. 1095).
9. The first National Park in the United States was Yellowstone, founded in 1872. The National Park Service under the Department of Interior was established in 1916.
10. Muir is arguably the most renowned and influential American naturalist and early advocate for wilderness preservation. He was founder of the Sierra Club.
11. 43 U.S.C. §§ 315-316o, June 28, 1934, as amended 1936, 1938, 1939, 1942, 1947, 1948, 1954 and 1976, see http://wildlifelaw.unm.edu/fedbook/taylorgr.html (accessed July 24, 2009).
12. Pub. L. 91-190, 42 U.S.C. 4321-4347, January 1, 1970, as amended by Pub. L. 94-52, July 3, 1975, Pub. L. 94-83, August 9, 1975, and Pub. L. 97-258, § 4(b), Sept. 13, 1982
13. 16 U.S.C. 1531-1544, 87 Stat. 884, as amended—Public Law 93-205, approved December 28, 1973, repealed the Endangered Species Conservation Act of December 5, 1969 (P.L. 91-135, 83 Stat. 275). The 1969 Act had amended the Endangered Species Preservation Act of October 15, 1966 (P.L. 89-669, 80 Stat. 926).
14. 16 U.S.C. §§ 1600-1614, August 17, 1974, as amended 1976, 1978, 1980, 1981, 1983, 1985, 1988 and 1990.
15. P.L. 94-579, 90 Stat. 2743. This law was in part to amend the deficiencies of the Taylor Grazing Act without repealing it.
16. See, R. M. Cawley, *Federal Land, Western Anger: The Sagebrush Rebellion and Environmental Politics*, Lawrence, KS: University of Kansas Press, 1993.
17. Cary Coglianese and Laurie Allen, "Does Consensus Make Common Sense? An Analysis of the EPA's Common Sense Initiative."
18. R. Edward Grumbine, "What Is Ecosystem Management?" *Conservation Biology*, 8 (1), 1994, pp. 27-38.
19. Gill Shepherd, "The Ecosystem Approach: Five Steps to Implementation," *Ecosystem Management Series* No. 3. 2004, IUCN—The World Conservation Union. http://app.iucn.org/dbtw-wpd/edocs/CEM-003.pdf.
20. Edward P. Weber, "A New Vanguard for the Environment: Grass-Roots Ecosystem management as a New Environmental Movement," *Society and Natural Resources*, 13, 2000, pp. 237-259.
21. Peter A. Walker and Patrick T. Hurley, "Collaboration Derailed: The Politics of 'Community-Based' Resource Management in Nevada County," *Society and Natural Resources*, 17, 2004, pp. 735-751.
22. Su Rolle, "The Applegate Partnership," Gary K. Meffe and C. Ronald Carroll, eds., *Principles in Conservation Biology*, 1st ed., Sunderland, MA: Sinauer Associates, 1997, pp. 612-616.
23. See, Edward P. Weber, "A New Vanguard for the Environment: Grass-Roots Ecosystem management as a New Environmental Movement."
24. National Research Council, *Environmental Issues in Pacific Northwest Forest Management,* Committee on Environmental Issues in Pacific Northwest Forest Management, Board on Biology, Washington D.C.: National Academies Press, 2000.
25. Lita P. Buttolph, William Kay, Susan Charnley, Cassandra Moseley, and Ellen M. Donoghue, *Northwest Forest Plan—the First 10 Years: Socioeconomic Monitoring of the Olympic National Forest and Three Local Communities*, General Technical Report

PNW-GTR-679, Pacific Northwest Research Station, Forest Service, U.S. Department of Agriculture, Portland, OR: 2006.
26. Julia M. Wondolleck and Steven L. Yafee, *Making Collaborations Work*, Washington, D.C.: Island Press, 2000.
27. Will Harmon and Matthew McKinney, *The Western Confluence: A Guide to Governing Natural Resources*, Washington, D. C.: Island Press, 2004.
28. Irene Frentz, Donald Voth, Sam Burns, and Charles Sperry, "Forest Service-Community Relationship Building: Recommendations," *Society and Natural Resources*, 13, 2000), pp. 549-566.
29. See, Edward P. Weber, "A New Vanguard for the Environment: Grass-Roots Ecosystem management as a New Environmental Movement."
30. See, Alexander Conley and Margaret A. Moote, "Evaluating Collaborative Natural Resource Management," *Society and Natural Resources,* 16, 2003, pp.371-386; Edward Weber, "A New Vanguard for the Environment: Grass-Roots Ecosystem management as a New Environmental Movement"; Steven Yaffee and Julia Wondolleck, "Collaborative Ecosystem Management Planning Processes in the United States: Evolution and Challenges," *Environment*, 3 (22), 2003, pp. 55-72.
31. Chrissy Coughlin, Merrick Hoben, Dirk Manskopf, and Shannon Quesada, *A Systematic Assessment of Collaborative Resource Management Partnerships,* A Master's Project completed for the School of Natural Resources & Environment, University of Michigan, 1999, at http://www.snre.umich.edu/ecomgt/pubs/crmp.htm#pdf (accessed July 24, 2009).
32. Montana Watershed Coordinating Council, 2008, at http://mwcc.montana.edu/groups /details.asp?groupID=42 (accessed March 20, 2009).
33. Perry Backus, "Timber in Transition: Blackfoot Challenge a Lasting Model for Conservation," at http://www.missoulian.com/articles/2007/02/07/news/top/news01.txt (accessed July 24, 2009).
34. *Blackfoot Challenge,* at http://www.blackfootchallenge.org/
35. *A Community-Based Approach to Watershed Protection and Management*, Henry's Fork Watershed Council, 2008, at http://www.henrysfork.com/watershedcouncil/ council.htm (accessed June 20, 2009).
36. Information from Partnership Resource Center, Henry's Fork Watershed Council, 2004, at http://www.partnershipresourcecenter.org/contact_us.php (accessed June 20, 2009).
37. 2008 statement by Henry's Fork Watershed Council, quoted from its website, http://www.henrysfork.com/watershedcouncil/council.htm (accessed June 23, 2009).
38. *Overview*, at http://www.qlg.org/pub/contents/overview.htm (accessed July 24, 2009).
39. Michael Hibbard and Jeremy Madsen, "Environmental Resistance to Place-Based Collaboration in the U.S. West," *Society and Natural Resources*, 16, 2003, pp. 703-718.
40. Jane Braxton-Little, "The Quincy Library Group," *American Forestry*, 10,1995, pp. 22-26.
41. Pub. L. 105–277, div. A, § 101(e) [title IV], Oct. 21, 1998, 112 Stat. 2681–231, 2681–305, as amended by Pub. L. 107–171, title VI, § 6201(d)(5), May 13, 2002, 116 Stat. 419; Pub. L. 110–161, div. F, title IV, § 434, Dec. 26, 2007, 121 Stat. 2153.
42. *Quincy Library Group*, at http://feinstein.senate.gov/qlg.html (accessed July 15, 2009).
43. See Michael Hibbard and Jeremy Madsen, "Environmental Resistance to Place-Based Collaboration in the U.S. West."
44. Information from CRCC's website at http://www.carangeland.org (accessed July 12, 2009).

45. See Michael Hibbard and Jeremy Madsen, "Environmental Resistance to Place-Based Collaboration in the U.S. West."
46. See, Michael Hibbard and Jeremy Madsen, "Environmental Resistance to Place-Based Collaboration in the U.S. West."
47. See, Peter A. Walker and Patrick T. Hurley, "Collaboration Derailed: The Politics of 'Community-Based" Resource Management in Nevada County."
48. See, Steven Yaffe and Julia Wondolleck, "Collaborative Ecosystem Management Planning Processes in the United States: Evolution and Challenges."
49. John Loomis, "Grass Roots Federal Land Management: Is this Devolution really the Solution?" *Western Economics Forum,* 1(1), 2002, pp. 1-4.
50. Renee Irvin and John Stansbury, "Citizen Participation in Decision Making: Is it Worth the Effort?" *Public Administration Review*, 64 (1), 2004, pp.55-63.
51. Stephen McCool and Kathleen Guthrie, "Mapping the Dimensions of Successful Public Participation in Messy Natural Resources Management Situations," *Society and Natural Resources*, 14, 2001, pp. 309-323.
52. The imperative of conservation in Great West Development was emphasized in the Chinese government policy from the beginning. See, Announcement by the State Council on the Policies and Measures in Implementing Great West Development [Guowuyuan guanyu shishi xibudakaifa ruogan zhengce cuoshi de tongzhi], State Council Issue (2000) Number 33.
53. The Chinese Communist Party reiterated the imperative of conservation and environmental health in its latest agricultural policy. See, "Resolution on Key Issues about Carrying forward Rural Reform and Development by the Central Committee of the Chinese Communist Party" [Zhonggong zhongyang guanyu tuijin nongcun gaige fazhan ruogan zhongda wenti de jueding], October 12, 2008.
54. Hu Angang and Wang Shaoguang, *The Chinese Economy in Crisis: State Capacity and Tax Reform (Studies on Contemporary China*, Watertown, MA: East Gate Book, 2001.
55. *Chinese and Hong Kong Nongovernmental Organizations*, The Wilson Center, 2007, at http://www.wilsoncenter.org/topics/pubs/Chinesengos07.pdf (accessed July 23, 2009).
56. Chinese name: Ziran zhiyou (Friends of Nature). More information at its website: http://www.fon.org.cn/channal.php?cid=774 (accessed July 26, 2009). As an example of its activities, see, "Friend of Nature Fights to Raise Awareness of Green Issues," South China Morning Post, page 14, March 11, 2005.
57. The environmental history of China demonstrated paradoxical human-environmental interactions, as Mark Elvin demonstrated in his book, *The Retreat of the Elephants. An Environmental History of China* (New Haven: Yale University Press, 2004). Chinese approach to conservation differed from that of contemporary Western civilization in that modification of nature was a common practice, but adopting sustainable approach for economic activities was the foremost concern.
58. Li Wenhua, *Agro-Ecological Farming Systems in China*, New York: The Parthenon Publishing Group, Inc., 2001.
59. The organization's website: http://www.snowland-great-rivers.org/index2.htm.
60. Tony Banks, Camille Richard, Li Ping, and Yan Zhaoli, Community-Based Grassland Management in Western China: Rationale, Pilot Project Experience, and Policy Implications," *Mountain Research and Development*, 23 (2), 2003, pp. 132–140.
61. Xu Jianchu and David Melick, "Towards Community-Driven Conservation in Southwest China: Reconciling State and Local Perceptions," ICRAF (International Centre for Research in Agroforestry) Working Paper Number 52, The World Agroforestry Centre, 2007.

Chapter 3

Values Conflict in American Environmental Federalism, with Implications for China

Sheldon Gen

INTRODUCTION

In the history of the United States, westward expansion of the early 1800s, followed by industrialization in the late 1800s, provided the foundations for the phenomenal economic growth it experienced through the 1900s. They also brought with them natural resource depletion and severe environmental damage. This path of initial strong economic growth coupled with environmental degradation has been a common experience of nations as they develop their economies.[1] Still, these painful environmental experiences potentially offer other nations cautionary tales of policy approaches to avoid during economic development. Specifically, as China pursues the economic development of its western regions, can the experiences of the United States help it avoid the environmental problems Americans experienced?

To answer this question, this chapter begins by broadly reviewing the history of American public policy toward the environment throughout its economic development. In the process, it reveals an evolution in the policies' values for the environment and their relationships with values for economic growth. It then turns to the contemporary United States, where the federalist structure of government has provided a natural experiment comparing the approaches and efficacies of environmental protection from the national level on one hand and the provincial and local levels on the other, each one representing different sets of values toward the environment. The analysis suggests that provincial/local advocacy and leadership, with its focus on environmental sustainability, appear to offer a more promising approach to environmental protection than the national approach has to date. The chapter closes with an argument of the cultural applicability of this approach to China.

THE EVOLUTION OF ENVIRONMENTAL POLICY IN THE U.S.

In order to identify the lessons that might be gleaned from the United States environmental experiences, we must first survey its evolution of environmental policy. And while the *protection* of the environment has manifested itself in U.S. policy only since the 1970s, the United States has had a long history of policies directed toward the environment.

The Era of Limitless Resources

The first era of U.S. environmental policy was one of environmental exploitation for the purposes of land development and economic growth. Through the 1800s, Americans viewed the environment as either a limitless resource for economic development, or an obstacle to development.[2] On one hand, during early economic development the environment provided seemingly limitless natural resource inputs for production, and limitless sinks for disposal of wastes and pollutants. Forests provided fuels, building materials, and paper products. Rivers were viewed as sources of potential energy as well as a natural system to carry away wastes from industrial processes. On the other hand, the environment was also seen as an obstacle to development, an unruly venue that had to be tamed for human utility. Wetlands were filled in order to develop the underlying land. Rivers were dammed to control flooding and provide steady water sources for irrigation and other developmental purposes. Mountains were carved to extract their rich deposits of minerals and to make smoother routes for transportation. These views of the environment were possible, even reasonable, at the time because the environmental impacts of these activities were still relatively small. With a national population of 5.3 million in 1800, and only 76 million one hundred years later, the spatial range of the environmental impacts were limited. Furthermore, the low state of technologies made these activities very labor intensive, further limiting the spatial and temporal environmental impacts of them. However, these views of the environment persisted in the United States through the industrial revolution and into the 1900s, and eventually resulted in environmental overuse and its severe degradation.

The ethical value reflected in these views is an anthropocentric individualism that prescribes the use of the environment to promote human welfare. White famously attributed this value for the environment to the Judeo-Christian ethic.[3] He argued that Christianity brought a linear perspective of time, in contrast to the earlier Greco-Roman cyclical view of time. In this linear progression, there is progress toward an end, and humans serve a divine purpose in this progress. The environment, in this view, is to be used for human's progress, and humans are to "subdue" it and have "dominion" over it.[4] While contemporaries have disagreed with this interpretation of Judeo-Christian teachings,[5] White's thesis was a description of the popular view during the U.S.'s early development. And nowhere was this view more clearly present in the policy of that time than in the concept of *manifest destiny*. The phrase reflected a nationalistic view that Americans are destined to bring their way of life to the world, and it justified the nation's westward expansion and development of the 1800s.

Preservation versus Conservation

As the exploitation of the environment persisted through the Industrial Revolution, the cumulative impact of human activity became apparent, and it was seen as unsustainable. This increased scope of impact was due primarily to the growth in population in the United States and the phenomenal technological advances that, together, made the speed and magnitude of natural resource use accelerate.[6] In reaction to this, beginning in the late 1800s, two broad approaches vied to replace dominion over nature with a more sustainable alternative.[7] One represented a direct refutation of dominance, promoted a more romantic view of nature,[8] and sought to preserve natural places for their own sake. John Muir, the founder of the Sierra Club, championed *preservationism*, and was instrumental in convincing President Theodore Roosevelt to establish a national park system. Preservationism is clearly biocentric, as it claims natural places have inherent value that is not dependent upon human utility, and thus they should be preserved.

The second approach that developed was an evolutionary adaptation of dominion over nature, taking into account the newfound scarcity of natural resources. *Conservationism* maintained the utilitarian view of the environment, but sought to manage its use to optimize the yield of benefits.[9] This anthropocentric holism still values the environment for its utility to humans, yet recognizes humanity's dependence on the environment. With industrialization and the growing population, no longer could the environmental impacts of economic development be ignored. Under conservationism, economic growth had to be planned in ways that ensured the sustainability of inputs and sinks from the environment. Natural resource management and agricultural economics are contemporary manifestations of this approach to environmental protection.

Contemporary Evolution of Conservation

Elements of both preservationism and conservationism remain in U.S. environmental policy today, but it seems clear that the values reflected in preservation have taken a second priority to those in conservation. This is due largely to the fact that preservationism, while successful in maintaining environmental conditions, does not allow for many utilitarian uses of the environment. Aside from recreational uses, preserved places are mostly off limits to development and resource extraction. In contrast, conservation offers a spectrum of policy approaches that integrate—to varying degrees—environmental protection and natural resource use.

The continued evolution of conservationism was spurred by the contemporary environmental movement beginning in the 1960s. The post-World War II population and economic booms accentuated the U.S.'s strain on the environment just when its people became more concerned with the quality of the environment. Rachel Carson published the now classic *Silent Spring*,[10] poetically describing the negative externalities of the nation's production practices. Donella Meadows and her colleagues published their alarming book *The Limits of Growth*,[11] delineating the natural resource limitations of unchecked economic growth, and gloomily predicting the collapse of both industrial capacity and population if the practices of resource exploitation persisted. These recognitions of impending environmental disasters eventually led to a set of strong environmental protection policies in the United States beginning in the 1970s. Most of these early national environmental statutes used regulations to limit the pollutants expelled by industries (e.g., the Clean Air Amendments of 1970, the Federal Water Pollution Control Act of 1972), or control the depositions of wastes and hazards (e.g., the Resource Conservation and Recovery Act of 1976, the Toxic Substances Control Act of 1976). Their collective purpose was to control and manage industry's impact on the environment, in order to protect human health and maintain an environmental condition that can sustain the economy.

But it is clear that the view of the environment as an anthropocentric resource persists in these policies. For example, the Federal Water Pollution Control Act requires—among other things—states to designate the human health and welfare uses of its bodies of water. Also, toxic substances are recognized as having utility in industrial production, so they are not banned by the Toxic Substances Control Act, only controlled to minimize environmental harm. These policies have become more sophisticated over the last thirty years, relying less on regulatory tools and more on economic incentive tools that still aim to meet standards for environmental quality while giving industry much more flexibility in how they can meet the targets. For example, pollution credit markets (e.g., carbon dioxide markets) and environmental banking systems (e.g., mitigation banking, wetlands banking) reflect the commoditization of environmental amenities and disamenities as a means to manage environmental impacts. In effect, these efforts internalize the environmental externalities of industrial activities. To facilitate this commoditization, environmental valuation techniques have been developed to assign monetary values to environmental amenities and disamenities.[12] This practice has become widespread in the United States, seeing support in numerous public policies that set environmental standards and establish monetary liabilities in environmental accidents.[13]

The creation of these disamenity and amenity markets illustrate how biocentric views of the environment in the West are secondary to the anthropocentric views. Airshed markets allow competing industries to trade the limited number of pollutant permits, so that such polluting activities can be made more efficient. Wetlands banking allows parties that preserve or restore wetlands to sell their credits to others who would fill other wetlands for development. While these advances in environmental policy have certainly helped curb the overuse of natural resources, they co-opt values for the environment into a utilitarian paradigm. Nonetheless, the broader, holistic view of the environmental context reflected in contemporary anthropocentrism is a vast improvement over the earlier individualistic view that led to unsustainable environmental degradation.

Conflicts in Values

This brief survey of U.S. environmental policy illustrates how environmental values in America have changed over the centuries, from a hostile dominance over nature to current sustainability approaches (See Table 3.1). Over the same span, however, economic values have remained relatively stable, and have consistently pursued growth. Even when economic activities were shown to produce significant damage to the environment, the solution could not infringe upon the economic imperative. Instead, environmental concerns were internalized into market transactions; environmental valuation tools were developed and widely used; market forces were created to build incentives for environmental friendly behaviors. Essentially, environmental amenities and concerns were commoditized to fit the economic paradigm of growth, and sovereignty in environmental policy remained with the consumer. And while this approach has recently contributed to improved environmental conditions, it is unknown how much better the condition of the environment might be, and how much more sustainable the economy might be, had earlier public policies taken a more holistic view of human dependence on the environment, or a more biocentric view to protect the environment.

Table 3.1 U.S. Environmental Policies and Their Underlying Environmental Values

Policy classification	Examples	Environmental value
▪ Dominion over nature	▪ Unlimited natural resource extraction; unlimited waste streams to the environment	▪ Anthropocentric individualism
▪ Preservationism	▪ Protected areas (wildlife refuges, national parks, etc.); endangered species act	▪ Biocentrism
▪ Conservationism	▪ Regulated natural resource use; pollution standards	▪ Anthropocentric holism
▪ Market-based conservationism	▪ Disamenity markets; mitigation banking; pollutant fees	▪ Anthropocentric holism

A TEST OF VALUES: ENVIRONMENTAL FEDERALISM

In the first eight years of the 2000s, however, glimpses of alternative approaches have been seen in the United States President George W. Bush sought to scale back environmental protections in favor of industrial activities and economic growth, just as the American public became increasingly aware of environmental problems, especially global warming. In the absence of national leadership to protect the environment, West Coast provincial and local governments took up leadership roles, and the potential outcomes are promising.

In the United States federal structure of government, policy responsibilities are distributed among national and provincial authorities. For example, the responsibility for national defense obviously lies primarily with the national government, while that for education traditionally lies with provincial and local governments. In issues of environmental protection, a compelling argument for national government leadership has been the transboundary nature of environmental problems. Because environmental problems often span municipal and even provincial jurisdictions (e.g., airshed and watershed problems), the broader geographic authority of the national government is often needed to coordinate, and coerce, the cooperation of the more local jurisdictions. Without national leadership, the argument goes, the affected localities may end up in one of two situations unfriendly to the environment: either a *race to the bottom*, with localities competing amongst each other for lower environmental standards in order to attract industries; or the *free-rider problem*, with individual jurisdictions shirking their environmental responsibilities under the assumption that the other jurisdictions will not. This often leads to the conclusion that the national government should indeed lead environmental protection because so many environmental problems are transboundary from a province's or a city's perspective. Indeed the national government of the United States has, since the 1970s, led all major policy efforts to protect the environment. In this environmental federalism, the national government sets policy and the state governments implement them.[14] The structure essentially defines a working relationship, rather than, for example, competing power centers or antagonists in environmental policy.

However, in the 1980s federal environmental protection waned under President Ronald Regan's economic policies, which favored an unregulated free market. During his terms as president, he cut the budget of the EPA and weakened environmental regulations.[15] This approach returned beginning in 2000 with the election of President George W. Bush. Bush's environmental policies heavily favored the pursuit of economic growth, as evidenced by the loosening of environmental regulation, early refusal to recognize (and thus regulate) anthropogenic causes of global warming, and the continued depletion of natural resources at unsustainable rates. This leadership vacuum in environmental protection at the national level left states and localities with opportunities to pursue their own environmental policies independently, even against the geographic challenges of environmental protection from the perspective of local jurisdictions. In effect, the retreat of national leadership in environmental protection policy under a federalist structure of government created a natural experiment of sorts to test the assumption that provincial and local governments will revert to anthropocentric individualism. In the absence of leadership from the national government, do provincial and local governments abandon environmental agendas in lieu of economic ones as hypothesized above, or do they rise to champion environmental values?

Recent policy activities in the California and other sub-national jurisdictions clearly demonstrate how provincial and local governments sometimes take leadership of environmental protection in the absence of national efforts. This despite the problematic economic incentives described above (i.e., the race to the bottom and the free-rider problem).

- *California's "Green Chemistry" initiative.* The national Toxic Substances Control Act governs the use of toxic substances, and sets standards for acceptable risks and exposures to such substances. However, it has not controlled the development of toxic substances used in production, and consumers are largely unaware of toxic substances associated with the products they buy. In response, California legislature and agencies have begun an initiative that will require every product to reveal its environmental footprint,[16] comprehensive measures of the environmental impact of the product's production, transportation, and disposal. The assumption behind the initiative is that consumers are not only interested in economic costs and benefits of the products they purchase; they are also interested in the environmental impacts of the products, and may select products—even higher priced products—that are less harmful to the environment.

- *California's vehicle emissions standards.* Under the George W. Bush administration, the national government had been reluctant to recognize the anthropogenic causes of climate change, and decidedly opposed to the regulation of greenhouse gas emissions from industrial and vehicular sources. The California government, on the other hand, not only recognized the causes of climate change, but also set ambitious goals to curb greenhouse gas emissions. It has proposed to reduce and regulate vehicular carbon dioxide emissions, by effectively requiring more vehicles sold in the state to be more fuel efficient.[17] Because California represents such a large portion of the U.S. market for vehicles, a required increase in fuel efficiency there would effectively result in industry-wide changes. The Bush Administration sought to block California's effort to essentially set national standards, but the national judiciary confirmed the state's right to set regulations on pollutants that exceed national standards.

- *U.S. cities' agreement to Kyoto Protocol targets.* Under Presidents Bill Clinton and George W. Bush, the U.S. failed to join the international community in signing the Kyoto Protocol, which set international targets for reducing greenhouse gas emissions. In the absence of national leadership, over 700 mayors of U.S. cities have signed an agreement to meet or exceed the goals of the Kyoto Protocol.[18]

- *Regionally led carbon policy.* Again, the absence of national leadership on curbing greenhouse gas emissions led sub-national governments to take action on their own. Four interprovince efforts are underway to set standards for carbon dioxide emissions under cap-and-trade programs. "Nearly everyone involved, from both the public and private sectors, would rather see a more coherent national carbon policy. But in the absence of [national] leadership, the [provinces] have gone to work on their own."[19]

John describes this local leadership on environmental actions as *civic environmentalism.* "The central idea animating civic environmentalism," he writes, "is that in some cases, communities and [provinces] will organize on their own to protect the environment, without being forced to do so by the [national] government . . . Civic environmentalism is fundamentally a bottom-up approach to environmental protection."[20] It relies less on the regulation of behavior as favored in national environmental policies, and more on grass-roots participation. John finds that provinces and localities are willing to spend their own budgets on environmental issues, not just national government's allocations, and they will enact their own environmental policies often going beyond national government requirements in environmental protection. In essence, he finds that the race-to-the-bottom argument for Federal leadership has not really materialized. Instead, states are often setting stricter standards.

Elazar's study of political cultures in the United States[21] may explain why civic environmentalism appeared in the western United States, and reveals the values underlying it. He described two major American political cultures relevant to this discussion. First, an *individualistic* political culture claims that the government should serve utilitarian purposes, especially economic ones. It should have minimal intrusion into private spheres like business, family, and church, except to keep the economy working well. This culture took geographic roots in the mid-Atlantic provinces and parts of the Great Lakes region. The values in this culture appear to support the anthropocentric individualistic view of the environment found in U.S. policies from independence through the contemporary environmental movement. In contrast, the *moralistic* political culture describes politics as a public activity centered on promoting the public interest. It prescribes widespread participation in political processes—as seen in townhall meetings, citizens' advisory councils, public debates, etc.—such that the public can influence policy making, and policies can reflect public values. This culture took root in the New England provinces, but has dominated the western provinces as well.

This geographic distribution of vying political cultures may partly explain why civic environmentalism is now appearing in California and the western region in general. In a culture that promotes local political activity and control, we would expect to see provinces and municipalities fill the leadership vacuum in environmental policy left by the national government. Furthermore, the culture's focus on public interests—as opposed to individual private interests—appears to explain why the western region has taken a holistic environmental stance in its policies. As Sagoff and Stone have extolled, in environmental issues people think and behave as citizens rather than as consumers.[22] Responsibilities to social and environmental conditions take priority over individualistic self-benefit. Indeed, Kempton, Boster, and Hartley found in their empirical study of stakeholders in environmental issues—ranging from environmental activists to sawmill workers—that Americans broadly reject a utilitarian approach to environmental policy.[23] Instead, the U.S. public is far more unified around environmental values than often assumed.

RELEVANCE TO CHINA'S CURRENT EXPERIENCE IN ENVIRONMENTAL POLICY

The historic American path to economic development has been environmentally unsustainable, a lesson that they are only now internalizing, requiring painful changes in its systems of production and habits of consumption. Economists and policy analysts have described this path of economic development as a common pattern in the West. The environmental Kuznets curve describes economic development's relationship with environmental condition as non-linear, but predictable.[24] During initial economic growth of a country, environmental degradation increases with income. This is due in part to the combination of the country's initial preferences for economic growth over environmental protection, and the country's initial view of the environment as a resource for economic growth. As the wealth of the country increases, however, and the marginal benefits of an ever-expanding economy decreases, environmental degradation also decreases. This is due to the new middle-class's concern for the environment and their willingness to engage economic activities in environmentally friendly manners. Post-materialism theory supports the Kuznets curve hypothesis by describing a two-rung hierarchy of societal goals beginning with economic growth, then advancing to environmental restoration and protection.[25] Like the Kuznets curve, post-materialism theory describes environmental degradation as a common path toward initial economic expansion. Indeed, these ideas have been used to explain environmental and economic concerns in the United States and throughout western and eastern Europe. Palmer, for example, reported empirical evidence showing that in countries where unemployment is high, environmental concerns are placed aside in favor of job creation.[26]

While these explanations suggest that environmental problems will work themselves out after economic growth has taken root, a conflict of values persists. In policies that commoditize environmental amenities, sovereignty still lies with the consumer, while they ignore the public's roles as citizens. As citizens the public has much different preferences than as consumers,[27] and such preferences express values for communal development, not just personal prosperity. Indeed, consumption is a weak measure of growth and development when community and environment are involved.

Thus, as China continues its rapid economic growth, we must ask whether these theories *must* describe China's developmental path. Must China experience severe environmental degradation in order to meet economic goals? Or can China lead the world by example of developing a strong economic base supporting its population without the Western pre-requisite of environmental degradation? The Brundtland Commission certainly believes a path distinct from the industrialized West's is not only possible, but necessary.[28] In its description of sustainable development, it claims that the economy and the environment are complements of each other, such that the long-term health of the economy is dependent upon the health of the environment. Several scholars have argued for this view of complementarity.[29]

An initial inspection of China's recent economic policies suggests that it is heading on the same path of the Kuznets curve as the U.S. Especially since the post-1978 reform era, economic growth has been pursued at the expense of the environment.[30] In the early 1980s, the central state gradually retreated from command of the economy, relinquishing land use and environmental decision to the market and consumers. The market replaced collective action, and has accentuated the environmental degradation. However, deeper analysis of the cultural contexts in China identifies factors that give high hope that it can champion environmental protection even as it pursues its economic goals.

The most important factor is the often cited *collectivist culture* of China, which values interpersonal cooperation to achieve common, social advancement. This culture lies in stark contrast to the individualistic culture of the United States, which values individual self-determination, and competition among individuals to achieve self-benefit. The American culture is clearly consistent with the utilitarian policies they have adopted to pursue economic growth and, only recently, protect the environment. Collectivism in China, however, is much more consistent with goals to protect the environment because of the public nature of so many environmental issues (i.e., exhibiting characteristics of public goods, club goods, or the commons; and environmental externalities of economic activities).

This Chinese cultural context, while sometimes over-simplified or presented as static, does exhibit itself in contemporary reality. Deng, Walker, and Swinnerton recently compared cultural values of Chinese with Canadians and indeed measured a significant difference.[31] They found that Chinese are more supportive of social-altruistic values related to the environment than are Anglo-Canadians. Such evidence has been recently supported by several studies of the differences between Asian and Western perceptions of the physical environment and the world.[32] These studies generally contrast a holistic perception held by Asians against a focused perception held by Westerners (See Table 3.2).

Interestingly, Chua, Boland, and Nisbett supported these perceptual differences with physical measurements of eye movement over different scenes and objects.[33] They found that Chinese and American participants have different viewing patterns of scenes, with Chinese paying more attention to the backgrounds and Americans paying more attention to the focal objects. That is, Chinese and Americans literally see the world differently, which explains in part their differences in perceptions of the world. Several argue these differences in perception stem from differences in social structures that emphasize scenes and relationships (Asian) versus objects and characteristics (Western).[34] For example, it is rare for Asian babies to sleep by themselves, whereas Western babies sleep in their own beds, or even in their own rooms. Also Mandarin-speaking mothers produce more verbs than nouns when speaking with their toddlers, while English-speaking mothers produce more nouns than verbs.[35] The former focuses more on actions and relationships among objects, while the latter focuses more on the objects themselves. It can also be argued that these perceptions also promote differences in behaviors and values that emphasize social advancement (Asian) versus individual advancement (Western).

Table 3.2 Western versus East Asian Perceptions[36]

Western perception	East Asian perception
▪ attend to a focal object in their view, analyzing its attributes; causal attributions tend to focus exclusively on the focal object without broader concern for its environment, and are therefore often mistaken	▪ attend to the broad scene of a view, noticing relationships among objects; causal attributions consider the context.
▪ live in less confining social worlds and can focus on attending to themselves and their goals	▪ live in complex social networks with prescribed roles and relations; thus attending to these relationships is critical to functioning
▪ Ancient Greeks saw matter as the composition of discrete objects/atoms; thought is fundamentally analytic; lived with fewer social relationships, valued autonomy and individuallism	▪ Ancient Chinese see actions as occurring in a field of forces among objects; see matter as interpenetrating substances rather than discrete objects; thought is fundamentally holistic; had relatively little sense of personal agency
▪ In conflicts Americans more likely to take sides, favoring one side while rejecting the other	▪ In conflicts, socially interdependent, "trying to coordinate their actions with those of others while minimizing social friction."[37]

Second in the factors that facilitate environmental protection is the high value that Chinese place on the environment, just as found in contemporary America.[38] Chinese culture has a long tradition of an ecocentric or biospheric worldview—supported by Taoism, Confucianism, and Buddhism—in which humans live in harmony with nature. This, combined with the collectivist culture that tends to subordinate individual interests in pursuit of group interests, suggest a very high potential for meeting environmental goals through cooperative, collective means.

Finally, China has already benefited from the experiences of the West and internalized these lessons in their own development. Heggelund and Backer, for example, explored the extent to which United Nations environmental bodies affect China's actions on environmental issues, and generally conclude that they do.[39] Christensen, Lisheng, and Painter go further by explaining when we can expect to find international learning among nations. When a country considers the adoption of reforms, they argue, it applies a compatibility test with its historical and cultural traditions.[40] High compatibility leads to higher acceptance and low compatibility leads to rejection, resistance, or slow adaptation to the cultural traditions. Applied to China, the first two factors identified above strongly suggest that pro-environmental lessons and goals from the West would have high congruence with Chinese culture.

CONCLUSION

This analysis of the American environmental experience, and the values underlying it, concludes that the anthropocentric, utilitarian view of the environment set the United States upon a path of economic development that led it through severe environmental degradation before the more recent policy efforts to restore and protect the environment. However, it also finds that cultural contexts in China suggests that it might avoid the environmental degradation of the Western path by embracing its values that are already aligned with the environment, rather than adopting the individualistic values dominant in the West.

The analysis leaves one remaining question, however: which level of government ought to lead environmental protection in China? In the U.S. experience, the national leadership has held on to an individualistic and utilitarian approach that has not met the public's demand for environmental protection. And in the absence of national leadership, the federal structure in the U.S. allowed provincial and local governments to step in. In China, however, the unitary structure of government may not allow for any leadership outside of the central government. Does this present any problem to the pursuit of environmental protection? Perhaps not. Byrne found that pollution growth is higher in a decentralized economy, while a centralized economy is optimal for pollution control during economic growth.[41] This is consistent with observations of economic transformations in Eastern Europe described by Baumgartl,[42] and quite relevant to China's political structure.

Still, if the central government delegated environmental authorities to provincial or local governments, evidence from the U.S. experience suggest that there may be benefits to environmental protection (See Table 3.3).

Table 3.3 Local versus National Government Approaches to Environmental Protection in the United States

Local government leadership	National government leadership
▪ managing the commons through cooperation	▪ controlling the environment through regulations, competition
▪ local communal values; citizen sovereignty	▪ consumer sovereignty
▪ driven by sense of place	▪ driven by individualism, self-interest (because no sense of place)
▪ examples: local land trusts, local government land use planning	▪ examples: regulations on pollutants, disamenity markets

Provinces and localities do have advantages over the national government in leading environmental policy making.[43] Among them, they are not dependent upon regulatory policy tools, while the national government is dependent upon it for uniformity and relative simplicity in implementation. Second, they are also smaller and less diverse than the nation, which allow them to tailor solutions to local circumstances and preferences, rather than finding general solutions to be applied broadly. Third, they can directly engage the citizens on a level the national government simply cannot. And finally, local leadership takes advantage of the sense of place in which environmental values are based on a commitment to place.[44] Together, these factors allow local communities to express their values for the environment from the perspective of the commons, valuing collective action and cooperation, instead of individualism and competition for the commons as found in more regional approaches to environmental protection (e.g., environmental disamenity markets). This is especially applicable to China as it takes advantage of the Chinese cultural views on interdependence, relationships between individuals and the surroundings, and community. Thus, while the central government is certainly in the position of authority to lead environmental protection in China, there may be significant advantages to delegating that authority to localities. Local leadership facilitates the direct participation of citizens in environmental protection, and this in turn offers the best hope for sustainability.[45]

NOTES

1. Rothman, D.S. & de Bruyn, S.M. (1998). Probing into the environmental Kuznets curve hypothesis. *Ecological Economics,* 25: 143-145.
2. Cutter, S. & Renwick, W. (2004). Environmental history in politics and decision making. In *Exploitation, Conservation, Preservation*. J. Wiley, New Jersey. Ch. 3, p. 36-60.
3 . White, L. (1967). The historical roots of our ecological crisis. *Science,* 155(3767):1203-1207.
4. Genesis 1:28, King James Version.
5. For example, see Dobel, P. (1977). The Judeo-Christian stewardship attitude toward nature. *The Christian Century* (October 12).
6. Gore, A. (1992). *Earth in the Balance: Ecology and the Human Spirit.* Houghton Mifflin Company, New York.
7. Norton, B. (1991). *Toward Unity Among Environmentalists.* Oxford University Press, Oxford.
8. A romantic view of nature is classically exemplified in Thoreau, H.D. (1854). *Walden; or, Life in the Woods.* Ticknor and Fields, Boston.
9. Norton, 1991.
10. Carson, R. (1962). *Silent Spring.* Houghton Mifflin Company, New York.
11. Meadows, D., Meadows, D., Randers, J., & Behrens, W. (1972). *The Limits to Growth: A Report for the Club of Rome's Project on the Predicament of Mankind.* Universe Books, New York.
12. For a thorough presentation of environmental valuation techniques, see Freeman, A.M. (1993). *The Measurement of Environmental and Resource Values.* Resources for the Future, Washington, D.C..
13. Gen, S. (2004). Meta-analysis of environmental valuation studies. Dissertation research. School of Public Policy, Georgia Institute of Technology, Atlanta, GA.
14. Scheberle, D. (1997). *Federalism and Environmental Policy: Trust and the Politics of Implementation.* Washington, D.C.: Georgetown University Press.
15. John, D. (1994). *Civic Environmentalism: Alternatives to Regulation in States and Communities.* Washington, DC: Congressional Quarterly Press.
16. Roosevelt, M. (2008). California officials launch 'Green Chemistry' initiative. *Los Angeles Times,* December 17. Retrieved on December 17, 2008 at http://latimes.com/news/local/la-me-green-chemistry17-2008dec17,0,7702359.story.
17. Healy, J.R. (2007). California wins right to regulate vehicle's emissions. *USA Today,* December 12. Yi, M. & Buchanan, W. (2009). Emissions rule waiver expected this spring. *San Francisco Chronicle,* January 27, p. A-1.
18. Healy, 2007.
19. Arrandale, T. (2008). Carbon goes to market. *Governing,* 21 (12): 26-30.
20. John, 1994, page 7.
21. Elazar, D.J. (1993). *The American Mosaic: The Impact of Space, Time, and Culture on American Politics.* Boulder, CO: Westview Press.
22. Sagoff, M. (1981). At the shrine of Our Lady of Fatima, or why political questions are not all economic. *Arizona Law Review, 23*: 1283-1298. Stone, D. (2002). *Policy Paradox: The Art of Political Decision Making.* Revised edition. W.W. Norton & Company, New York, NY.
23. Kempton, W., Boster, J., & Hartley, J. (1995). *Environmental Values in American Culture.* MIT Press, Cambridge, MA.
24. Rothman & de Bruyn, 1998.

25. Inglehart, R. (1971). The silent revolution in Europe: intergenerational change in post-industrial societies. *The American Political Science Review, 65*(4): 991-1017. Inglehart, R. (1981). Post-materialism in an environment of insecurity. *The American Political Science Review, 75*(4): 880-900.
26. Palmer, R. (1991). Economic diversification and the environment in a small economy. In David McKee, ed. *Energy, Environment and Public Policy: Issues for 1990's*. Praeger, New York. Pages 157-165.
27. Sagoff, 1981; Stone, 2002.
28. World Commission on Environment and Development (1987). *From One Earth to One World: an Overview by the World Commission on Environment and Development.* World Commission on Environment and Development.
29. For example, see: DeCanio, S.J. (1997). Economic modeling and the false tradeoff between environmental protection and economic growth. *Contemporary Economic Policy,* 15: 10-27. Kassiola, J.J. (2003). "The 'tragedy' of modernity: how environmental limits and the environmental crisis produce the need for postmodern values and institutions." In J.J. Kassiola, ed., *Explorations in Environmental Political Theory: Thinking About What We Value.* M.E. Sharpe, Armonk, New York. Pages 14-36. Templet, P.H. & Farber, S. (1994). Complementarity between environmental and economic risk: an empirical analysis. *Ecological Economics,* 9 (2): 153-165.
30. Muldavin, J. (2000). The paradoxes of environmental policy and resource management in reform-era China. Economic Geography, 76(3): 244-271.
31. Deng, J., Walker, G.J., & Swinnerton, G. (2006). A comparison of environmental values and attitudes between Chinese in Canada and Anglo-Canadians. *Environmental Behavior,* 38 (1): 22-47.
32. For example, see: Nisbett, R.E. & Masuda, T. (2003). Culture and point of view. *Proceedings of the National Academy of Sciences,* 100 (19): 11163-11170. Nisbett, R.E. & Miyamoto, Y. (2005). The influence of culture: holistic versus analytic perception. *Trends in Cognitive Sciences,* 9 (10): 467-473.
33. Chua, H.F., Boland, J.E., & Nisbett, R.E. (2005). Cultural variation in eye movements during scene perception. *Proceedings of the National Academy of Sciences,* 102 (35): 12629-12633.
34. Miyamoto, Y., Nisbett, R.E., & Masuda, T. (2006). Culture and the physical environment. *Psychological Science,* 17 (2): 113-119. Nisbett & Masuda, 2003.
35. Nisbett & Miyamoto, 2005.
36. Sources: Nisbett & Masuda, 2003; Nisbett & Miyamoto, 2005.
37. Nisbett & Masuda, 2003, p. 11163.
38. Deng, Walker, & Swinnerton, 2006.
39. Heggelund, G. & Backer, E.B. (2007). China and UN environmental policy: institutional growth, learning and implementation. *International Environmental Agreements, 7*: 415-438.
40. Christensen, T., Lisheng, D., & Painter, M. (2008). Administrative reform in China's central government—how much 'learning from the West'? *International Review of Administrative Sciences,* 74 (3): 351-371.
41. Byrne, M.M. (1997). Is growth a dirty word? Pollution, abatement and endogenous growth. *Journal of Development Economics,* 54(2, December): 261-284.
42. Baumgartl, B. (1993). Environmental protest as a vehicle for transition: the case of Ekoglasnost in Bulgaria. In A. Vari and P. Tamas, eds. *Environmental and Democratic Transition.* Kluwer Academic Publishers, Netherlands. Pages 157-175.
43. John, 1994.
44. Norton, B. & Hannon, B. (1997). Environmental values: a place based theory. *Environmental Ethics,* 19: 227-245.

45 . For strong arguments for local leadership in sustainable development, see: Gundersen, A. (1995). *The Environmental Promise of Democratic Deliberation*. University of Wisconsin Press, Madison WI. Prugh, T., Costanza, R., & Daly, H. (2000). *The Local Politics of Global Sustainability*. Island Press, Washington, D.C.

Chapter 4

Urban Growth Control and Farmland Preservation in West China

Xiao Hang Liu

INTRODUCTION

Urbanization is accelerating worldwide. For the first time in human history, half of the world's total population will live in urban areas by the end of 2008 according to the United Nations. In China, the speed of its urbanization is unprecedented. In 1990, the percentage of population residing in cities and towns was only 26 percent. By 2001, this number has increased to 36 percent and by 2006 it reached 44percent.[1] As China's industrialization continues to expand rapidly and population grows at an annual rate of 0.63 percent, it is expected that China's urban population will reach 50 percent by 2015 and even 70 percent by 2035.[2]

Much of China's urbanization took place in the coastal area where the outcome of the open policy and the economic reforms has been felt most strongly. In contrast, urbanization in west China is still under development in the sense that the west has significantly less number of cities and the population of these cities is relatively smaller. However, this situation is changing rapidly. As the labor-intensive agriculture sector is being modernized, each year there are 12 to 15 million surplus labors flood into the non-farm job market.[3] In order to absorb these surplus labors as well as to reduce the economic disparities between the east and west China, the Chinese government initiated an effort in the late 1990s to speed up the industrialization in the west. One of the key strategies employed is to increase urbanization in the west so that cities become more capable of absorbing the rural labors migrating to the urban area. As a result, seven new cities were built in the west between 1998 and 2002, increasing its urban built-up area 17 percent. Land acquired by the government for development purpose was 90.35 km^2 in 1998, 67.21 km^2 in 1999, 105.8 km^2 in 2000, 1090.81 km^2 in 2001, and 1524.18 km^2 in 2002. The nearly 17 fold increase from 1998 to 2002 suggests that the acceleration of urbanization speed is apparent, and is likely to continue into the near future.[4]

Experience on urbanization worldwide shows that urbanization is typically accompanied by the loss of cropland and open space. For example, Japan lost 52% of its farmland between 1955 and 1994 and South Korea lost 48 percent of its farmland from 1965 and 1994. China is not immune to this problem. According to China Land Resource Yearbooks,[5] the total decrease of cultivated land in China in 1997 was only 461,400 ha. In 2002, the decrease became 2,027,400 ha. As urbanization in the west catches up with that in the east cost, the west becomes the main region where farmland loss happens. In fact, 83 percent of cropland loss in China in 2003 took place in the west. Cropland, especially those with high quality agricultural soils and convenient access to water, is a very important resource in the west. According to an agriculture census in 1996, cropland accounts for only 7.3 percent of the overall land use in the west, totally 49.59 square kilometers. Moreover, the majority of the cropland is rated as of medium or low productivity. Soil loss, severe water shortage, desertification, and geological hazards are the reasons behind the scarcity. Under such circumstances, the loss of prime farmland due to urbanization becomes especially concerning.

Farming satisfies the desire of nearby metropolitan residents for locally-grown produce, as well as provides open space and scenic vistas. Preserving existing high productivity farmland will also reduce the need to convert forests and ranges to farmland which is a main cause of increased desertification in the west. For historical reasons, most metropolitan regions in the west are located in areas with quality soil and easy access to water, e.g. alluvial plains, basins, or oases dotted on the diluvian fans along the feet of mountains and the rivers. Not surprisingly, these are also where prime farmland is located. If urbanization takes the form of urban sprawl, which is ubiquitous in China, a loss of prime farmland will be unavoidable.

Hence is the challenge on how to preserve farmland in the face of urbanization. In this chapter, the importance of farmland preservation and urban growth control in the west is first analyzed from a geographic perspective. Given that the United States has faced, and is still facing, similar challenges in its urbanization, we review the policy instruments that have been employed in the United States and the lessons learned. Finally, drawing on the U.S. experience on urbanization, a few suggestions are presented on how to increase urbanization in the west at a minimal price of farmland loss.

URBAN GROWTH AND FARMLAND LOSS IN WEST CHINA

West China is a term typically referring to the remote hinterland in China where ethnic minorities concentrate. Geographically, the west contains nine provinces or autonomous regions which can be grouped into two subregions: the Southwest which consists of Sichuan, Guizhou, Yunnan, and Tibet; and the Great Northwest which consists of Inner Mongolia, Shannxi, Ningxia, Gansu, and Xinjiang. Topographically, the Southwest is characterized by plateaus, basins, and a few large river systems. In contrast, the Northwest is known for its deserts and arid or semi-arid weather.

Because of the complex topography, cities in the west are mainly located in diluvian fans, alluvial plains, or oasis along the feet of mountains and the rivers. For example, Chengdu which is one of the two largest cities in the west with over two million population, is located in Chengdu plain at the bottom of Sichuan Basin. The other megacity whose population is also over two millions, Chongqi, is located at the edge of the Sichuan Basin. Other important cities in the west are also found in areas with relatively flat topography, fertile soil, and easy access to water. For example, Yinchuan in Ninxia is located at Yinchuan Plain, LanZhou in Gansu is located at Hetao Plain, Xi'an in Shannxi is located at Guanzhong Plain, and Urumqi (meaning "*beautiful pasture*") in XinJiang is located in Zhungeer basin. All of these cities are the capital cities of their provinces or autonomous regions, hence play an important role in the economic development of west China.

In 1999, the Chinese government launched the Go-West initiative which aims to speed up the industrialization of the west by increasing its urbanization level. As a result of this initiative, many cities have developed the plan to expand themselves significantly in the next few years. For example, Sichuan province plans to reach urbanization level of 38 percent by increasing its built-up area from 1201.41 km^2 in 2002 to 2776 km^2 in 2010, which is equivalent to double its urbanized area in just eight years. Chongqi also plans to double its size in the next 10 years by expanding the city towards the flat area in the north. Noticeably, the expansion plans are not limited to megacities only. Wuzhong, a middle sized city in Ninxia autonomic region, also plans to expand itself from 23 km^2 to 190 km^2. Growth plans of similar magnitude can be found in many other cities in the west.

Existing cities in the west are mainly located in areas with advantageous physical geography in terms of climate, soil quality and access to water. This is understandable as historically these cities were built when China's economy is dominated by agriculture. Fertile soil and convenient access to water resource enables high productivity of farming, which is necessary to sustain the large urban population. Farmland has always been a scarce resource in the west. Although the west boasts of 71.04 percent of the total area of China, its topography is dominated by plateaus, desserts, and mountains, most of which are not habitable. Only a few large plains and basins are of decent size to serve as an agriculture base. In fact, the percentage of plains and basins in the west is less than 10 percent. Among them, farmland occupies only 7.3 percent. To further aggravate the situation, much of the farmland has only medium to low productivity. Nationwide, two thirds of the China's farmland belongs to medium to low productivity. Among them, 75 percent are from the west. It can be seen that productivity of west farmland is much below the national average.

The fact that large metropolitan centers happen to be where prime farmland is found in the west creates a serious challenge to urbanization. Current urbanization in the west has mainly in the form of urban sprawl. Suburban and rural area in the fringe of metropolitan areas are often the first to be acquired and developed into built-up area. The direct consequence of this conversion is the loss of prime farmland which is often adjacent to large cities. Although this loss can be compensated by developing new farmlands in areas further away, it has to be noted that farmland increase in the west is often at the price of forest and ranges. The impact of such conversion to the fragile ecological system of the west can be detrimental, resulting in loss of soil, desertification, geological hazards, etc. More importantly, these newly gained farmlands are usually of much poor quality compared to the farmlands lost in the fringe of cities. Hence, even though the total area of farmland can remain the same or even increases during urbanization, there will still be a loss of overall productivity not to mention the environmental impacts induced.

U.S. POLICY INSTRUMENTS FOR URBAN GROWTH MANAGEMENT AND FARMLAND PROTECTION

Preservation of prime farmland in the face of urbanization is not a unique challenge in China. In recent years, the annual loss of farmland in the United States is about 1.3 million acres a year. Nearly half of this land is prime farmland and most of it is located in metropolitan counties where 80 percent of Americans live.[6] At an age when the value of farmland for raising crops, fruits, vegetables, and livestock is much less compared to the land for housing and commercial development, how to protect farmland from being converted or sold to development purposes is a serious concern in the United States too. In the past few decades, various public policy instruments have been developed to address this challenge with varying degrees of success. In this section, those instruments which are believed to be applicable to China's west are reviewed. Lessons learned from the U.S. experience are also discussed.

Overview of Public Policy Instruments

In the United States, the government intervenes urbanization and farmland protection by exercising three types of instruments: public acquisition, regulation and incentives.[7] Public acquisition is mainly exercised by the government for the collective benefits of the general public by purchasing lands. Since all lands are state owned in China, this method may not be necessary but can be employed in conjunction with master land use planning. Regulation refers to the mandatory approaches whose violation results in sanction or threats to sanction. In the U.S., regulation has been used to manage urban growth as well as farmland preservation. One example is development moratoria which halt the issuance of building permits in order to buy time to develop solutions to growth-generated problems. This approach was used by the city of Los Angeles in 1998 where a building moratorium on new large projects was extended for 10 and half months as an effort to update the city's aged general plan.[8] The rate of growth control is another example of regulation which puts an upper limit on the number of building permits issued annually according to the existing or projected capability of public facilities. Urban containment policies regulate urban growth by drawing urban growth boundaries (UGB), urban service boundaries, or Greenbelts. All cities in the state of Oregon are required to delineate its UGB. The Portland metropolitan area is especially often examined. Finally, planning mandates is a regulation which develops comprehensive plans to guide land-use decisions. In the United States, all local governments in the western states require land use planning. The regional and state level mandate can assure that the comprehensive plan at the local level is consistent with state plans for highways, sewers, transit, airports, and regional parks.[9]

Incentive approaches resort to monetary credits or materials resources to change the land use decisions. For example, the government may promote compact or infill development by providing government subsidies or reducing development impact fee. The city of Austin in Texas waives the development fees in desired development zones in order to encourage infill development. Alternatively, a municipality can also encourage high-intensive land uses (e.g., apartments) by using split-rate property tax, i.e. a higher-tax rate on land values and a lower-rate for buildings. Studies show that this approach has been successful in Pennsylvania to revitalize inner cities.

In addition to urban growth management, regulation and incentives are also available to protect farmland and open space. A key regulatory approach is zoning, or more precisely, large-lot agricultural protection zoning, which has been widely employed at the local and state levels across the U.S. to protect farmland. Experiences show that counties successful in farmland preservation share the key feature of adopting agricultural zoning ordinance which allow no more than one dwelling per 25 acres[10]. Marin county and Sonoma county in California, which are located in the San Francisco Bay Area Metropolitan Region, are the leading U.S. counties in farmland preservation. Marine county has the agricultural zoning of 1 house per 60 acres, while the zoning standard in Sonoma county varies.[11] A large number of incentive-based policy instruments have also been developed to protect farmland. An example is the right-to-farm law. When residential development encroaches into rural areas, homeowners may complain or even file lawsuits against farmers because of the noise, smell and other unpleasant impacts from farming. The right-to-farm law protects farmers and ranchers from such lawsuits, thus creating an incentive of farming[12] (Bradbury, 1986). Nearly every state in the U.S. has at least one right-to-farm law, which is stronger in some local governments than the others.

Successes and Lessons

The public policy instruments on urban growth control and farmland preservation had varying degrees of success in the United States. A few key lessons have emerged.

Lesson 1: It is very important to develop a strategy for urban growth control and farmland preservation, instead of simply to protect whatever available. Policy instruments such as the master land use plan and agricultural zoning need to be in place to indicate where farmland should be preserved and where developments are permitted. This lesson applies to jurisdictions at various levels. In the United States, the Natural Resource Conservation Service (NRCS) which supervises the federal farmland preservation program has been criticized for a lack of preservation strategy. By allocating funds to private land trusts which may have a land use decision different from that by public land use planning, there is the concern that there will be inconsistent efforts on farmland preservation.

Lesson 2: Public instruments on urban growth management and farmland preservation should be employed collectively in order to achieve the best results. Experiences have shown that a package of complementary reinforcing policy instruments is much more effective than single policy instrument. An example of success is Oregon which uses exclusive farm use (EFU) districts to preserve farmland for farming in the long run. At the same time, Oregon uses urban growth boundaries (UGB) to prevent urban sprawl; exurban districts to accommodate the demand for rural residential lifestyles without harming commercial farm operations; and the right-to-farm laws to create incentives to keep farming, and comprehensive plans to legitimize the entire package.[13]

Lesson 3: Implementation of policy instruments is critical to make the instruments work. Otherwise, no matter how effective the policy is, the intended outcome may not be achieved. For the purpose of effective implementation, the coordination between federal, state, and local governments as well as between neighborhoods and with the stakeholders is very important. Effective coordination helps to avoid the circumstance where one jurisdiction achieves its goal in urban growth control or farmland protection at the price of the other jurisdictions. A lesson is from Boulder in Colorado which had used greenbelt to contain urban growth only to spur development leapfrogging over the greenbelt and spill over to the nearby communities.[14]

Lesson 4: Regulations and incentives are only one approach to achieve the goal of sustainable urban development. Stakeholder participation in the decision making process has been proven equally important. By conveying information to the general public through campaigns, educational workshops, or townhall meetings, concerns of all participation groups are more likely to be addressed. Calvert County in Maryland used this approach to support downzoning. A set of built-out scenarios were developed and distributed to residents to acquire public opinion. Eventually, residents favored a scenario which saves farmland and money significantly by reducing the density to 50 percent. Similar successes have been reported by other counties in Maryland.

PRESERVING PRIME FARMLAND IN THE FACE OF URBANIZATION IN WEST CHINA

Urban Sprawl versus Smart Growth

In current west China and even the entire China in general, urbanization has been implemented mainly in the form of urban sprawl, resulting in a drastic circumstance where the area of a city is much larger than its actual built-up area. However, urbanization is not equivalent to urban sprawl. By definition, urbanization is the process to increase the proportion of the population residing in communities with urban characteristics. This suggests that the goal of urbanization is not to create megacities by expanding urban boundaries, nor it aims to convert rural or suburban areas completely to cities. Instead, the purpose is to mitigate the urban-rural dichotomy so as to reduce the disparities between them. From this perspective, sprawl is not the appropriate avenue to advance urbanization in the west. Smart growth, which values long-range regional sustainability, advocates compact development, public transportation, mixed land use, a range of housing choices etc., should be adopted.[15]

Development and Implementation of a Comprehensive Land Use Plan

A comprehensive land use plan and farmland protection strategy provides a guidance to urbanization. It is based on scientific research and is an outcome of coordinating the needs of various stakeholders, hence the plan and the strategy should be enforced. Learning from the U.S. experience, a clear statement on the strategy including who is to implement it and how, what are the objectives and how will the objectives be evaluated are ought to be addressed. The incorporation of an assessment plan is especially important, as Bengston et al. pointed it out as one of the main lessons learned in the U.S. experience.

Development of Small Cities

Cities in the west mainly occupy plains and alluvial fans. The vast hilly land is yet to be taken advantage of. Hilly areas may not suitable for high productivity farming. However, their geological structure determined that they may possess natural resources which can be exploited by the surplus labors in the rural area. Some cities in China are built on hills such as Qingdao and Chongqing. By learning from these cities, it is possible to convert the lands not suitable for farming to development. For example, the natural gas resources in Xinjiang make it possible to establish micro cities in areas distant from oasis. Small cities can also be built to host the industries moved out large metropolitan areas during rezoning and redevelopment process. In this way, not only the existing land of large cities is utilized more effectively, the environmental impact of the industries can also be mitigated. This avenue has great potential if the government can assure that sufficient utility service is available, such as sewage, electricity, transportation, hazard prevention etc.

Utilizing Urban Land Use Information System

Up-to-date knowledge on the inventory of urban land and prime farmland is the starting point to develop a scientific strategy to control urban growth as well as preserving farmland. Given the rapid urbanization in the west, advanced spatial information technology such as Geographical Information Systems (GIS), remote sensing, and Global Positioning Systems (GPS) should be employed. An urban land use information system is an essential information infrastructure. With its capability in data storage, display, and analysis, it is possible for the decision makers to monitor and visualize the dynamic change in land use. Recent advancement in geospatial technology such as Internet-based GIS and high resolution satellite remote sensing opens the opportunity to higher public participation in the land use decision making process.

CONCLUSION

The vast surplus labors in the rural area of west China have important implications to China's sustainable development in the future. To absorb these labors into the non-farming job market, urbanization has been pointed out as an important strategy on the grounds that it will speed up industrialization. Yet urbanization is often accompanied by the loss of prime farmland. It is therefore imperative to develop mechanisms for effective urban management. A wide range of important issues are closely tied to growth management policies, including environmental conservation, quality of life, taxpayer protection, efficient urban form, transportation planning, urban revitalization, environmental justice, affordable housing, and others. We hope the policy instruments reviewed in this chapter will foster the understanding of urbanization, and the lessons learned in the United States will help design more effective growth management programs in China.

NOTES

1. NBS: China's total population shrinks to 56% of total. Available online: http://news.xinhuanet.com/english/2007-10/22/content_6925292.htm
2. United Nations Population Fund: State of World Population 2007. Available online: http://www.unfpa.org/swp/2007/english/chapter_1/different_speeds.html
3. Johnson, Gale, 2002. Can agricultural labour adjustment occur primarily through creation of rural non-farm jobs in China? *Urban Studies* (2002), 39: 2153-2174.

4. Zhang, Q., 2004. Reconsideration on cultivated land conservation during urbanization in the west. Journal of Chongqing University of Industry and Commerce.
5. China Land Resource Yearbooks, 1997, 1998, 1999, 2000, 2001, 2002.
6. National Resources Conservation Service, 2001. National Resource Inventory, 1997 (revised). Washington, D.C.: U.S. Department of Agriculture.
7. Bengston, D., Fletcher, J., and Nelson, K., 2004. Public policies for managing urban growth and protecting open space: Policy instruments and lessons learned in the United States. *Landscape and Urban Planning*, 69, 271-286.
8. Los Angeles Times, 1998. City Extends Development Moratorium by 10 1/2 Months, June 24.
9. See Johnson, Gale, 2002.
10. Nelson, A., 1992. Preserving prime farmland in the face of urbanization: Lessons from Oregon. *American Planning Association Journal of the American Planning Association*, 58(4), 467-488.
11. Daniels, T. and Bowers, D., 1997. Holding our ground: Protecting America's Farms and Farmland. Island Press, Washington D.C., p.352.
12. Bradbury, D., 1986. Agricultural law: suburban sprawl and the right to farm. *Washburn Law Journal*, 22, 448-468.
13. Daniels, T., 2004. *Farmland preservation policies in the United States: Successes and Shortcomings*. Farmland Preservation Conference: Protecting Farmland for Farmers, Guelph, Ontario, Canada, June 2004.
14. Pendall, R., Martin, J., and Fulton, W., 2002. Holding the line: urban containment in the United States. Discussion paper. Center on Urban and Metropolitan Policy, the Brookings Institution, Washington D.C. (available online: http://www.pewcenter.org/doingcj/research/r_ST2000.html).
15. Environmental Protection Agency, 2009. About Smart Growth. Available online at http://www.epa.gov/smartgrowth/about_sg.htm

Chapter 5

Archaeological Site Management as Environmental Policy in the United States and China: Implications for Western Expansion in China

Gary W. Pahl

China and the U.S. have independently worked to preserve their heritage manifested in archaeological resources and arrived at parallel but unique practices to effect protective control for their cultural resources. This chapter offers a comparative look at how these distinct policies have developed and potentially reveals directions China might contemplate as it experiences a growth surge into its western regions.

This chapter draws together the historical sequence of policies in the American experience with cultural resource management dating to the 1880s and its uneven legal modifications in response to changing national and local interests. America's conflicting issues down to the unsolved questions surrounding legal ownership of heritage resources are summarized.

In comparison, China's current rigorous federal top-down approach to cultural resource management is seen as effective and might best avoid proliferation of the multiple American layers of conflicting local, state, and national laws which cloud clear solutions.

ARCHAEOLOGICAL SITE MANAGEMENT IN THE UNITED STATES AND CHINA

The remains of historic and prehistoric cultures in the U.S. and China exist as fragments of the past embedded in the contemporary environment stratigraphically layered back into antiquity and preserving the sequential passage of cultural events. The physical impact of ancestral cultures also affects the environment spatially across regions and sites reflecting chronological and qualitative changes of culture over horizontal expanses. The archaeological site formation processes as well as its collapse and abandonment processes are muddled together in contemporary environmental expanses, usually buried into and beneath the modern surface of the environment. These cultural remains remind us that the apparent natural state of contemporary environments is actually a mélange of the natural biotic, hydraulic, geographic, and geologic interactions as well as the remnant admixture of behavioral remains of previous cultures scattered throughout it.[1]

The history of American interest in recovering and preserving its prehistoric and historic antiquities are relatively young compared to that of China. An early serious investigator and preserver of American cultural resources is none other than the famous statesman, Thomas Jefferson (1743-1826), who is appropriately identified academic archaeological circles as the father of American Archaeology. Jefferson excavated out of curiosity on his own property in Monticello, Virginia in the late 1700s. He stratigraphically uncovered the remains of ancient Native American settlements and carefully retained and studied them for his own pleasure as, what we now call, an antiquarian. He published his findings in 1787 in an American edition of *Notes on the State of Virginia*. Jefferson had collected extensive information on local Native American linguistics and developed his own hypothesis as to the origin and development of the native population preceding European invasion. Archaeology recognizes Jefferson's contributions to Archaeology but has matured well beyond the antiquarian collecting of curiosities by Jefferson in colonial America.

Today U.S. government agencies, business interests and archaeologists are left with the unenviable task of making sense out of the natural and cultural resource mixture and recommending how modern development should be handled with regard to them. How should they be treated during plans for expansion and building? Should cultural resources simply be plowed under or given special consideration for the impacts which the development will have on the resources so that they might be preserved completely or in part ? Should archaeological site vestiges be completely preserved to the restriction of growth and development projects? Should individuals be allowed to take, claim or loot cultural remains from these sites and regions? Who owns the relics of past prehistoric and historic cultures? Who rightfully makes the ultimate decision for cultural historical environment and site management and impact during times of expansion and development?

As if these questions were not enough to make the handling and planning of environmental management enough of a puzzle, there has also been a recent epiphany or awakening in American society to the realization that minority group descendants claiming relation to ancestral inhabitants of cultural resources or sites may wish to make their concerns and opinions know about the disposition of cultural resources and remains affected by modern development and expansion plans. This issue has come to a head and grown in importance over the last 20 years or so in America. It is worth looking at the relatively brief development of American thoughts and legal reactions about managing its cultural resources juxtaposed to China's several thousand years of consideration and care of its antiquities.

American Historical Experience with Cultural Resource Management (CRM)

America's own westward expansion exhibits a relatively slow response to cultural resource protection and management as part of an environmental protection process. Initial laws were passed to protect archaeological sites from looting and export of artifacts near the beginning of the 1900's. These legal responses were followed by laws generated over half a century later to identify cultural resource sites on a wider regional basis and provide some form of regulation for alteration and excavation of such sites. Only recently has there been a legal response to attend to the concerns of groups affected by development in cultural resource bearing environments and modification of sites in them.

Generally speaking, American environmental protection policy has only tangentially dealt with cultural resources in an effective way. American environmental and cultural resource management law reflects on-going difficulty in preservation and management of sites in developing environments even with intended avoidance of looting, negative construction impacts, and total destruction. These difficulties are compounded by our own expansion into natural/cultural resource environments with and building of roads, shopping centers, and homes as building costs are directly and legally tied to limitation of identification, recovery, or preservation of archaeological resources. Archaeological resource protection typically takes a back seat to development and is attached as a tertiary impact issue slowing down development.

It is interesting to look at our environmental laws and regulations intended to manage cultural resources to see how they have evolved and to observe how, in practice, they were generated for differing reasons and wind up failing to provide for careful cultural resource management which cannot be easily circumvented by those wishing that environmental and culture resources did not exist as an impediment. It is an interesting and incomplete saga.

American Antiquities Act

The first American legal attention to management of its antiquities in the environmental context occurred in 1906 with the national level Antiquities Act. The inspiration for the legislation arose with the realization that looting of American archaeological sites was resulting in the decimation of cultural sites and the theft of artifacts representing ancient native cultures. The guiding feature here was a desire to stop the looting and destruction of archaeological sites by individuals for personal gain or exportation. It was becoming apparent that archaeological resources were a non-renewable resource. Once a site was excavated it became understood that it cannot be dug again. Once a site was looted, the lost material remains could never be recovered again.

Consideration of minority descendants or their concerns was not an issue with the earliest legislation. Rather, the early protective legislation indicated a need to set aside and preserve archaeological sites as a unique record of the past. There was also a sense of need for some level of planned preservation of threatened and disappearing cultural remains. The early laws also attended to assessment and planning for projected protection of regions thought to contain cultural resources remaining as vestiges of unidentified ancient cultures. The positive impact resulting for the cultural resources and the surrounding or supporting environment was that it became illegal to dig on archaeological sites or public land without a legally drawn permit from the government. Only museums, universities and other scientific institutions were allowed to apply for such excavation. The president of the country in 1906 (Theodore Roosevelt) was invested with the authority to designate national monuments and landmarks for preservation. Ironically, this Act did not help save archaeological sites from development or even address the issue.

National Historic Preservation Act

With American expansion to stimulate a down economy through immense construction projects such as hydraulic dams for generation of electricity in the 1930s, archaeologists increasing became alerted to the need to bring legislation to bear on protecting cultural resources and their supporting environments not just from looting but from heavy development affecting large tracts of environment in the country which did not consider management or preservation of cultural resources. The National Historic Preservation Act NHPA finally followed the Antiquities Act in 1966 and put teeth and more specific details into specifically "mitigating the impact of development." The development intended by the government under NHPA was that which occurred on government held land, at the behest of government agencies or with government funding.

Ironically, the law also indicated that even excavation by trained archaeologists ultimately had an adverse impact on sites and site regions. Archaeologists in American responded with a mathematical sampling methodology whereby archaeological resources might be sampled in part but a substantial portion held in reserve for future generations of archaeologists with improved technology and methods. Under the most extreme preservationist position, only threat of looting, construction or erosion would warrant site excavation. This cautionary posture in the law, even for archaeologists, created a strong sampling theory application in American archaeology that persists today so that non-threatened sites are usually left alone and preserved. Other sites would have to be strategically sampled by a small percent in order to ensure preservation or continued future testing with more sophisticated methods and questions.

NHPA also provided for a legal mandate for an extensive inventory and survey of federal land for presence of cultural resources (The National Register of Historic Places). NHPA also called for new bureaucratic agencies to record the sites on their respective properties resulting in the Bureau of Land Management (BLM), National Parks, National Forests and so on. Interestingly no American agency has yet been able to identify all of the cultural resources on its federally held land even though the preponderance of American's archaeologists now work for BLM or kindred agencies. Over 90 % of land in the US is federally owned.

The National Register of Historic Places created by NHPA began the process of identifying what is now a growing, but still woefully incomplete, list of American cultural resources. Sites are significant enough to be registered on the list if they are associated with important pre-historic or historic events, important people in the past, or if the site exhibits distinctive traits of a style of architecture, construction method, or high artistic value. Similarly, if the site is known to contain data important for prehistory or history, it can be nominated, listed, and preserved. The concept of "significance" now has protective legal implications for archaeological sites. Sadly, even sites included on the National Register of Historic Places enjoy no complete protection from destruction.

American CRM and preservation archaeology is also concerned with "question-oriented archaeology" or archaeology guided by the precepts of scientific method. The reason for the question orientation is to create a situation in which the archaeologist must defend and demonstrate in detail the rationale, planning and outcomes behind excavating and "impacting" a site to the public and other professional archaeologists. It clearly makes for many conflicted and paranoid archaeologists in America but usually results in well reasoned cultural resource management and less wanton site destruction. Over 90% of archaeology conducted in America today is conducted by CRM archaeologists under the effect of legal compliance archaeology.

A sad result generated from all of the legal management of the sites and attendant environment is that no effective requirement exists for the management of resulting site data, collections. Each agency has its own hackneyed bureaucratic system for handling its reports and recovered material culture (artifacts). Similarly, each agency restricts public access to site and collection information through its own bureaucratic maze and in non-centralized locations. Reports on the collections and sites, if they exist at all after excavation or survey, are woefully lacking and not required or financially supported by the agencies in a way that is fully and easily available to the public let alone the academic community. Any reports or proposals are usually "canned" in their production and not well focused on scientific questions or detailed descriptive reporting. This is particularly sad and awkward at best in light of the current advanced state of scientific applications and technology available for archaeological and environmental studies today.

Archaeological Protection Act

Another valuable federal law, Archaeological Protection Act (ARPA). was put into effect in 1979. ARPA intensifies archaeological sites defense as well as archaeologically sensitive environments. ARPA re-enforces earlier laws and prohibits the sale or transport of artifacts acquired illegally from federally owned or affected sites. The penalties for such infractions increase to fines up to $250,000 with a possible five years in prison.

It is worth noting that privately held property can still escape the purview and long reach of the law in America. This creates a highly awkward site protection situation where private land owners can frequently mine their land at will for cultural resources unless public but non legal pressure is bought to bear for site protection of an archaeological site. Sites recorded on the National Register of Historic Places can also be modified. Even National Register sites on private property can still be excavated or eradicated at the whim of the property owner. National Register status merely provides a potential tool for making it more difficult to impact an archaeological site on the privately held property.

State Laws and Local Ordinances

Some states and counties have written and enforce their own highly strict local laws and ordinances which inhibit private owners from digging archaeological sites on their property. (California Environmental Quality Act or CEQA is a good state example). The laws, regulations and ordinances, which tightly control issuance of permits and inspections for any development, building or modification...even on private property, for say, a new tennis court or swimming pool, create an indirect way to legally protect cultural resources on privately held property. Interestingly but not surprising, local ordinances are usually written and controlled by local planning commissions composed of developers who are interested in seeing growth and construction go forward with reasonably little impediment. This is tantamount to having the proverbial "fox guard the chicken house" as far as archaeology and cultural resource preservation is concerned. The successful implementation of the proposed development in a timely manner is paramount in the perceptions of the planning commissions.

The catch in putting any of these legal proceedings to protect or conserve cultural resources into motion then is in the enforcement. If no one is willing to publicly protest a construction project impacting a cultural resource site and generate enough interest whereby legal intercession can be effected and paid for, the project and site impact can go forward. Financially backing such a complaint is usually the largest impediment to conserving cultural resources. The intent of the ordinance is supposedly to control the construction permit process on any property under their jurisdiction closely enough so that the public is informed of all of the ramifications and impacts of the project before permits are issued.

Required attention to environmental impacts include impacts to cultural resources and normally require that the company hired to conduct the environmental impact report for the project also hire a CRM archaeologist to review and report on any cultural resources which may be impacted by the project. The CRM archaeologist makes recommendations as to need for mitigation measures to sample, test, partially preserve, or fully preserve a cultural site from the impacts of the project. The costs incurred for the EIR and the mitigations recommended are borne by the project developer.

What ensues during such a development process where environmental and, by extension, cultural resources, are concerned is a tug-of-war between the "developer," the CRM specialist, and possibly "Native American Most Likely Descendants." The project developer who will impact the cultural resources with his project can be required to modify the construction project to avoid or lessen the impact to the site. The CRM archaeologist can be compelled to lessen the demands for mitigation to protect the site, requests for testing for clarification of significance of the site, and ultimately cut a deal with the developer to allow the project to go forward with some level of compromise to the cultural resource. The final outcome usually results in a compromise of the integrity of the archaeological site or cultural resource with the original intent of the law shoved to the sidelines.

Native American Graves Protection and Repatriation Act NAGPRA

The above laws, among others, are the American government's legal reaction recognizing the need to conserve and carefully manage its cultural resources in specific environments as historically important American records in the form of material culture. In 1990 the government was moved for the first time to side-step the subject of cultural resource law and move into a new realm of cultural sensitivity in America whereby the opinions of Native American minority groups were to be taken into consideration with regard to cultural resources and environmental law. More specifically, the government passed a law to protect the disturbance of human remains of Native American groups on federal and tribal lands and forbid sale of the remains. The instigation of the law arose originally at the behest of Native American Hawaiians under the aegis of Senator Inouye. The law, Native American Graves Protection and Repatriation Act (NAGPRA), gave Native American groups a voice in disposition of their ancestral cultural resources remains: human remains, funerary objects, sacred objects, and objects of special significance to the Native American group.

Issues in Contemporary American CRM Law

The current cultural resource management situation in America, with all of its historic layering of laws since 1906 protection and conservation of archaeological resources in America may be perceived from the outside as an effective system for their intended purpose. There are, however, some inherent problems with this rather byzantine complex of laws and ordinances which, for all their good intent to preserve and protect environments containing cultural resources. This phenomenon results in continued loss, mismanagement and neglect of cultural resources in the face of development and looting.

1) The laws and ordinances crossing from the most centralized federal levels to the smallest local town levels have loopholes of which the most determined developer and/or looter will take full advantage. Inevitably the financial burden of summoning legal argument and enforcement of the existing laws plays an important role in determining the survival or loss of cultural resources.

2) Individual agencies operating beneath the federal, state and local levels of government typically have their own bureaucratic requirements, regulations, and forms. Examples here are National Park Service, U.S. Forestry department, and State Historic Preservation Office under the aegis of the Department of the Interior the federal government. This phenomenon creates a heterarchy of competing legal requirements in a lateral conflicting sense rather than a clear hierarchical structure.

3) A full knowledge of the applicable available laws and ordinances is imperative in order to understand which laws might be applicable, under what circumstances and when. The level of knowledge is specialized to the point where it would be beneficial to be an environmental cultural resource management specialist and a practicing attorney at the same time when engaging in American CRM.

4) None of the laws has a broad enough reach to be applied across the country or in localities. Again, this results in situations where it is difficult to determine which laws or ordinances might be applied for use in protecting a particular site or case. Federal, state and local entities frequently come into conflict over which body or agency has priority or responsibility.

5) The complexity of the laws, regulations, and ordinances across so many levels of government and subrogated agencies provides excellent opportunities for devising legal defense of looting and destruction of cultural resources and ways around the legal provisions themselves.

6) The interpretation of the laws and regulations for application to cultural resource management and preservation are left to attorneys who characteristically have no cultural resource management or archaeological background.

7) Legitimate archaeological research with cultural resources is hampered by the layers of bureaucracy and legalese intended to protect archaeological sites. Academics have performed less and less archaeological research on sites since the 1990s because of these obfuscations.

8) Access, documentation and publication of data and material cultural remains are required by legislation. Accordingly, research and publication of information concerning the findings of surveys, excavations, and artifacts have been relegated to the "back burner" by government agency and CRM work on cultural resource sites.

Chinese Administration of Cultural Heritage

The history of China's care and involvement with its cultural heritage must be acknowledged for its extreme antiquity. Chinese traditional historiography and preservation of cultural relics is steeped in a moral purpose of studying the past as a repository of relevant ancestral experience and a confirmation of the Confucian era classics. This ancient moral backdrop in Chinese cultural heritage preservation is not characteristic of American archaeology and preservation law. It is, in fact, often a difficult concept to explain to American audiences.

As early as the Shang dynasty, extending from the 16th to the 11th Century B.C.E., tombs like that of Empress Fu Hao exhibited the reverence the Chinese held for their antiquities in preservation of a wide range of relics of jade and bronze from far-reaching corners of the Shang empire. It is fairly clear rom the archaeological record and suggestions from the classical literature of Zhou and Han that reverence for antiquity and ancestors in China more than likely harks back well into the Neolithic Period.

Empress Fu Hao's tomb included jades from Chinese unrecorded Neolithic antiquity, possibly as old as a thousand years by Fu Hao's time (1200 B.C.E.) at the Shang capital of Anyang. They include Neolithic pieces from Shijiahe culture in what is now Hubei Province, the Liangzhu culture of the Yangtze delta area, the Longshan culture of modern Shandong, and Hongshan culture of Liaoning Province. All of these ancient and distant heirlooms drawn together in the private funerary accoutrements of Fu Hao bespeak her recognition of extreme value in Neolithic antiquities. Undoubtedly these were personal possessions of Fu Hao and can be said to represent an early example of an attempt to preserve antiquity if not exotics.

Similarly, Lu Ta Lin from the much later Sung dynasty (960-1279) published what was intended to be an elitist bronze collectors guide, the Kao Gu Tu in A.D. 1092 and commenced what is formally recognized in China as an age of traditional antiquarianism. This again was an honoring of the past with the moral injunctions of the Classics. The Kao Gu Tu recorded bronze and jade artifacts dating back to Shang dynasty held in imperial and private collections.

It further represented an early Chinese industrious effort to closely analyze the ancient bronzes and their inscriptions with an eye for preservation and correct historical identification. This was a strong preservationist step in the right direction.

Today, state administration of China's cultural heritage operates at the behest of the federal level Cultural Relics Bureau and is administered by the State Administration of Cultural Heritage. The legal structure of cultural resource management is clear cut and unambiguous. The State Administration of Cultural Heritage sits at the top of the legal hierarchical pyramid and mandates all forms of possible cultural resource management in the country. The U.S. enjoys no such ultimate or overarching controlling legal entity for cultural resources.

The Law Guide for the State Administration of Cultural Heritage identifies the national and provincial level agencies that protect its cultural heritage sites. [2] It further controls all development and building construction, site repair, and excavation permits related to identified cultural resources. It essentially delegates all management of cultural resources to provincial offices under the State Heritage Department. Provincial authorities in turn review, hear and ultimately decide on impact issues or projects related to cultural resources. They identify, hear proposals, and decide on teams or units which will act on their decisions. The reporting hierarchy reflects a carefully structured scheme of responsibility ending with the national secretariat level.

An example might be recent CRM projects conducted in Chengdu, Sichuan Province in which a series of cultural heritage finds were uncovered during the process of road building on the edge of the city.[3] The Chengdu City Archaeological Team affiliated with the Archaeological Institute heard of the discovery and contacted provincial officials who in turn stopped the project and called upon the City Archaeological Team to recover the find under a permit issued by the provincial authorities. This was a neat and seamless operation with no jurisdictional confusion and the important recovery of a series of earthen sacrificial altars covered in elephant tusks, ceramic vessels, and a number of jade bi disks were all carefully unearthed and remanded to the research offices of the City Archaeological Team for professional interpretation and preservation.

While it is possible that the construction crew could have graded through the cultural relics, the fact that the legal structure is so firmly in place in China and that most of the players in such a process are well aware of their respective roles in such circumstances, everything fell fairly neatly into place without much drama. Because the cultural heritage system seems so well understood at various levels in Chinese society, the likelihood of destruction of important finds though jurisdiction confusion or bumbling appears greatly reduced. According to professional Chinese colleagues knowledgeable in such issues, legal punishment for the crime looting or cultural relic destruction is very stiff. Depending on the importance of the site damaged, such as an important tomb, looting can be punished in the courts with a two-year sentence to an ultimate penalty of execution. Effecting such punishment for archaeological resource damage or looting in the U.S. would be enhanced significantly with addition of more rigorous punishment along these lines.[4] The execution penalty may seem a bit draconian but an inveterate archaeologist might contemplate supporting it. As it stands, it is difficult to effect legal prosecution of any kind in the U.S. without great expenditure of energy and expense let alone draw such a cultural resource case into the courts.

Such a recovery of an archaeological site during a construction process would probably not be handled so smoothly in the U.S. Upon initial discovery of a relic or site, if an archaeological monitor was not available on site to identify it, the construction crew may simply have obliterated the site or pocketed the finds. If the finds had been reported to local authorities in the U.S. what followed would be a sorting out of what agency or level of legal authority should be recognized and contacted and the proper authority whose jurisdiction could move forward on deciding what to do with the find. The awkward process is usually laborious and costly for the developer and threatening to the cultural relics.

Considerations for CRM In Western Expansion and Development in China

1. The essential suggestion for managing cultural resources as part of growth and expansion is that the Chinese Cultural relics Bureau continue in it centralized hierarchical management of its cultural heritage as it exists. It appears to be a very efficient and clear process under one national lead agency with all final authority.

2. A Consideration which may already be in effect in China as suggested by this current meeting is that the Cultural Relics Bureau might begin extensive advanced planning including regional surveys to attempt identification of specific cultural resources and register them according to a national typology of cultural resources. Such regional surveys, if they do not already exist, could be assigned to local professional archaeology units or research teams which could register sites and report back to the provincial and national level. Essentially this would entail the creation of a log or record of all cultural resources regionally in intended growth areas anticipated for future development.

3. A recommendation that may run aground with the centralized government is open and concerted consultation with minority groups who may claim affinity to archaeological ruins and remains. This recommendation is akin to the NAGPRA legislation in the U.S. which has not been fully resolved or applied to date. The Chinese central government recognizes its minority populations but does not involve them fully in decisions about the modification and development of lands bearing their cultural resources. This is a serious human rights issue that China will inevitably have to grapple with if it wishes to fully appreciate the value of its cultural resources and its minority populations. The U.S. has certainly not perfected the application or interpretation of its own

NAGPRA legislation but it has made steps in the right direction by passing the law in the first place.

4. Most critical for consideration would be provision of government funds set aside exclusively for the regional surveys, and for the formal reporting of the findings. Funds should also be made available in advance and secured for mitigation of impacts by local professional archaeology units to allow recovery of important resources which might be slated for destruction or impact through eventual population expansion and development. In the U.S. many laws and regulations are passed but remain ineffective without earmarked funding set aside to address the suggested mitigation, preservation and management of identified cultural resources.[5]

NOTES

1. R. Bagley, "Shang Archaeology." In Michael Lowe and Edward L. Shaughnessy (eds.), *The Cambridge History of Ancient China* (Cambridge University Press, 1999), pp. 124-231.
2. Chinese State Cultural Relics Bureau, *Law Guide* (Chinese State Administration of Cultural Heritage, 2008).
3. D. Hurst and R.L. Kelly, *Archaeology*, 4th Edition (Thomson Wadsworth, 2006).
4. Interview with He Nu, professor in the Department of Archaeology at the Chinese Academy of Social Sciences on March 23, 2009.
5. L. Sebastian, "Archaeology and the Law," In J.R. Richman and M. P. Forsythe (eds.), *Legal Perspectives on Cultural Resources* (Altamira Press, 2004), pp. 3-16.

Chapter 6

Environmental Consciousness Change: A Comparative Study of the United States and China (1950-2008)

Qiu Zhong and Guoqing Shi*

INTRODUCTION

China encountered three major environmental issues at the beginning of the 20th century—the Yellow River drying of 1997, the Yangtze River flooding of 1998, and the Peking storm of 2000. Those events are symbolic of the ecological disasters that erupt from the accumulation of environmental destruction. At the same time, sociological researcher Shaokang Liu pointed out "these issues will probably provide an opportunity to accelerate the process of dealing with environmental problems in China."[1]

How do the aforementioned environmental issues become the turning point for policy change? Triggered by environmental concern change is possible. Dunlap, who is being credited as the co-founder of the field of environmental sociology, defined environmental concern as, "the degree to which people are aware of problems regarding the environment and support efforts to solve them and/or indicate a willingness to contribute personally to their solution." [2]

So, how does environmental concern come out of these events? Environmental disasters are a necessary but not sufficient condition for change as John Hannigan points out in his book, *Environmental Sociology: A Social Constructionist Perspective.* "Environmental problems do not materialize by themselves; rather, they must be 'constructed' by individuals or organizations who define pollution or some other objective condition as worrisome and seek to do something about it." For Hannigan the social institutions that most impact the evaluation of environmental events are those of science and mass media. According to Hannigan these institutions play a major role, "in constructing environmental risks, knowledge, crises and solutions." [3]

The author thinks that these two factors are more compatible with Western society than with Chinese society. However, emergent environmental problems have been shown in multiple arenas in both societies. China has her own characteristics, and the mainstream of Chinese environmental concern is focused on policy and economy. I want to give a comparative work in this article of the environmental consciousness between the United States and China. After the comparative work, I will discuss the influence of these different factors on environmental consciousness and use the discussion to draw out some conclusion from the comparative work.

LITERATURE REVIEW

The early work of Catton and Dunlap brought out the narrow anthro- pocentrism of classical sociology.[4] The publishing of their research proved to be a watershed moment in environmental sociology, refocusing research in the field and marking a new epoch. They thought that the traditional type of sociological research, modeled after Durkheim, neglected the nature of environmental influence on human society. They pointed out that the foundation of the traditional western world view is that human beings are the masters of the world, using the resources of the world to deal with current and developing problems. After the transition from a Sociology of Environment to an Environmental Sociology, they were challenged by HEP (human exemptionalism paradigm) to NEP (new ecological paradigm) contrasting traditional sociological thought and emerging environmental sociology. Later, Dunlap amended NEP and used it as the measure in the United States.

In *Environmental Sociology* Hannigan pointed out "in the early 1990s, environmental scholars were still by and large committed to a strong 'realist' position, whereby the gravity, configuration and causes of ecological damage and destruction were thought to be obvious. By contrast, many researchers recognized that environmental risks and knowledge were by no means self-evident, but rather the product of social definition and construction."[5]

Human beings need to follow the law of nature, but at the beginning of the new China establishment, the people and government still held the old world view. The leaders of China thought humans would triumph over nature.[6] So at that time, people in China believed human could rely on science and technology and thought that the Communist systems could overcome the laws of nature.

As environmental problems were exacerbated by social development, China has become increasingly aware of the environmental issues. Learning from the painful environment experiences of Western countries, China's environmental consciousness was impelled to the next step.

Compared to Western societies, China's environmental sociology discipline appeared late. In the middle of the 1980s-1990s, environmental sociology literature in China mainly included a brief introduction of environmental sociology, foreign environmental academic conferences and papers translated from abroad. Dayong Hong, the leader of Chinese environmental sociology research, distinguished two types of environmental (theory) research; the ecology environmental sociology and the sociological environmental sociology.[7]

Based on the Social movement theory, Dayong Hong discussed the opportunity of explaining environmental issues by social factors coupled with social transformations. He pointed out that structural transformation, institutional transformation and the change of social values are directly exacerbated by the deterioration of environmental conditions, and those transformations shaped characteristics of environmental issues of China. [8] Hong amended Dunlap's NEP scale for China, based on the 2003 Chinese General Social Survey data.[9] However, there has not been any research based on the amended scale in China.

The case study in the rural countryside of China opened another side of research in environmental sociology. Ajiang Chen, using this qualitative study, made it feasible to measure environmental issues on the local level avoiding the problem of distortion in quantity research.[10]

1950s-1960s

The United States: Budding and Awakening

At the end of the 19th century, the United States completed the process of modern industrialization. Total industrial output value began to exceed the total agricultural output value. At the end of the 1920s, the realization of electrification began. In the 1930s, the United States suffered a serious economic recession, along with natural disasters also occurred. Out of this crisis, President Franklin Roosevelt took a number of measures, one of which was resource protection legislation, providing work as a form of relief, such as planting trees and grass and building irrigation. The most outstanding effective measures were the comprehensive treatment of the Tennessee River Valley, the Great Plains dust governance of the United States, and the building of a series of water conservancy and hydropower projects in the western area.

In the 1950s, the invention of the transistor caused the United States another scientific and technological revolution. Keypad telephones, stereo radio, drying/washing machines, air conditioners and frozen food emerged at that time. The U.S. economy had experienced unprecedented prosperity. As the Americans enjoyed the wealth, a blemish of the industrial society appeared gradually. Water pollution, nuclear pollution, chemical pollution, and air and noise pollution caused by the automotive industry, followed the rapid economic development.

The water pollution control law, which was promulgated in 1948, marked the beginning of the federal government's involvement in environmental protection. However, the environmental pollution of the United States did not become a global issue at that time. The lack of the financial capacity of the federal government limited environmental protection in this period as well. Local governments took care of most of the management responsibilities and focused on the protection of water resources. In conclusion, the effects in different areas varied a lot.

In the 1960s, environmental degradation as well as environmental consciousness was generated among the elite of the United States. In 1962 *Silent Spring*, written by American marine biologist Rachel Carson, was published with a large number of facts. She demonstrated that industrial pollution damages the life on the earth including human beings and stated that the industrial revolution caused ecological damage and first alarmed the world to the seriousness of environmental protection. Carson said in the book: "For the first time in the history of the world, every human being is now subjected to contact with dangerous chemicals, from the moment of conception until death." [11] The book was widely read and made the public concerned about contamination. In 1972, Carson's work facilitated the ban of DDT in the United States.

China: No Consciousness

When the new China was just founded, the Chinese industrial foundation was very weak, the population was not too large and the environment was not a problem. At that time, environmental protection was only considered in keeping the work and living place clean and improving the outlook of urban and rural areas. These were the main contents of the patriotic public cleaning campaign. This situation continued until the mid-1950s. Due to the limited scale of production, environmental problems had no major effect on economic development. From the country's overall situation, the main environmental issue was the regional ecological damage and contamination. There was no true sense of the concept of environmental protection in this period; neither was progressing toward the development of related laws and regulations.[12]

After the 1950s, the environmental pollution of China became more and more serious. Anti-pollution campaigns appeared in some developed countries. In China, people had some preliminary understanding of environmental protection. However, most viewed environmental protection as only related to air and water pollution control, solid waste treatment, (called "three wastes" treatment) and the elimination of noise with technical measures and management. The main purpose was to eliminate pollution to protect human health.

In addition, since 1957, China experienced the political situation, (Great Leap Forward, Great Steel-Iron Making, and the People's Commune Movement) and the economic situation, (three years of natural disasters and food shortages) which made people strongly demand natural resources. However, there was no comprehensive understanding of the proper use of natural resources. Most people were neither familiar with the concept of environmental protection; nor understand the situation in this area. They thought Chinese environmental problems were not important and that they need not be too anxious. At the same time, most people under socialism did not admit there was a problem of environmental pollution. They thought of it was a product of capitalism and was an incurable disease of the capitalist countries. Whoever said that China had a pollution problem was negating the superiority of socialism.

Some people thought environmental issues were the same as health problems, over emphasizing the damage to human health by environmental pollution while minimizes the seriousness of environmental pollution's harm to society in all other areas.

In the science area, the book *Geographical Environment and Social Development*[13] published in China written by Ze Wu discussed the geography of social development within the Marxist perspective. At the same time in the 1950s, there was issued another work about the relationship between natural environment and human society *Geographical environment, population and social development*[14] written by Zhende Wang based on Wu's writing. Wang further added population variables to strengthen the understanding of the relationship between the geographical environment and social development.

1960s-1970s

The United States: Comprehensive Transformation

The study, Longitudinal Studies of Public Concern for Environmental Quality (1968-1976) shows that the peak was at the end of 1969 and the early part of 1970. [15] 22 USC Section 2151p-1, adopted by Congress through P.L. 87-195 Dec 30, 1969, as amended, provides for protection of tropical forests, by placing a high priority on "conservation and sustainable management of tropical forests" and to "the fullest extent feasible." With this basic environmental law, environmental protection came onto the legalization. Congress officially brought the United States Environmental Protection Agency (USEPA) into existence.

On April 22, 1970, an estimated 20 million people across the nation attended the environmental movement with a goal of a healthy, sustainable environment. [16] It was then celebrated as Earth Day and marked the birth of the modern environmental movement with the public concern for environmental quality reached unprecedented levels. This supportive public opinion provides a valuable resource for environmental movement. The environmental movement has had a great impact on environmental consciousness and environmental protection, even on socio-economic policy and the future of the movement will depend heavily on the degree to which environmentalists can effectively mobilize the support. [17]

In the seventies of the last century, the United States faced too many environmental pollution issues such as acid rain, nuclear pollution, dredged material pollution, oil pollution and so on. Because of these issues, President Nixon created the EPA in 1970. His plan was to establish the EPA for pollution control purposes, because "the environment must be perceived as a single, interrelated system." [18] From then on, the U.S. Congress passed many environmental bills tackling everything from air and water pollution and garbage to protection for fisheries, marine mammals and endangered species.

As environmental issues appeared on the front page of newspapers and the covers of magazines, words and phrases such as ecology, environmental costs, resource depletion, eutrophication of rivers, environmentalism and other similar words became popular even in several famous magazines, *Times, Fortune, Life, The New York Times, Washington Post* and so on.

The Club of Rome was founded in April, 1968. The Club raised considerable public attention in 1972 with its published report *The Limits to Growth. The Limits to Growth* is a book modeling the consequences of a rapidly growing world population and its finite resource supplies.[19] After that, "only one earth" became the slogan of the Stockholm Conference on the Human Environment UN.[20]

The trend data in Gallup surveys, especially covering the years from 1969 to 1970, shows the change in environmental consciousness. 17 percent of respondents selected pollution ranked it ninth among the ten top problems in 1965, while the 53 percent selecting it in 1970 placed it second only to crime reduction. [21]

China: Initiation Stage

The Cultural Revolution put the Chinese economy onto the verge of a collapse between 1966 and 1976.[22] Environmental problems have become more and more serious. Several environmental issues showed destruction and depredation at that time. In 1972 an emergency caused by pollution happened at Dalian Bay. The rising tide brought black water and ebb tide left the beach black. The pollution left more than 5000 Chinese acres (1 Chinese acre =666.6m^2) of dead shellfish on the beach. Every year China loses more than 10,000 kilograms of trepang, more than 100,000 kilograms of shellfish, and more than 150,000 kilograms of corbicula. Some fisherman even acquired Shui Yu (methyl mercury poisoning) sickness from eating those fish or shellfish.

With the advice of Premier Enlai Zhou, China sent one delegation to participate in the Stockholm Conference on June 5, 1972. This Human Environmental Conference pointed out that environmental problems are not just linked with technology but also have a close relationship with social-economics; these problems not only making the issues local but also global. This conference is recognized as the beginning of environmental concepts that promoted environmental protection in China. The senior leader of China paid much attention to environmental issues after this conference, and environment became significant along with social and economic development.

On May 20, 1973, the Chinese Council held the first national environmental protection meeting in Beijing. Protection policy increased in this meeting for purpose of planning and distributing protection of the environmental for the masses of people. On November 17, 1973, three state Commissions co-promulgated the "industrial emissions test standard." This "standard" provided the basic law for the Environmental Protection Agency, ended the unsystematic policy of Chinese environmental protection, and also made feasible environmental monitoring.

Dependence upon those policies influenced Chinese society; people began to understood the relationship between environment and development and realized that environmental protection was not only control, but also (or more importantly) the rational utilization of resources. Economic growth is limited by environmental capacity.

1970s-1980s

The United States: the Recoil after the Recession

In the early 1970s, after the peak of the environmental movement, early environmental protection policies made progress. As a result of sufficient funds and great attention by many people in the United States to the pollution problems, environmental protection methods were greatly enhanced, which further improved the coordinated capacity of measures in all levels of the environmental protection system. However, the huge expense of environmental protection in the United States became an important component of budget deficits. Strict control measures also caused a negative effect on the economy and this caused great concern for both the United States government and the public.

During the years of 1973 to 1975, the United States suffered a structural crisis which was characterized economic stagflation. At that period, the United States and the Western world had a serious economic crisis. The panic over a shortage of resources (especially oil) made environmental concerns take back stage in the United States. The economic issues had a more profound impact on people's lives than the environmental issues. In 1984, compared to 1980 before Ronald Reagan was elected, the laws and regulations published by the Government to intervene in social activities were reduced a lot, greatly easing the standards and regulations in environmental protection, technology security, health care and consumer protection that were published previously.

Nevertheless, after a brief setback in the 1970s and 1980s, the environmental concerns in the United States resumed. *Time* Magazine's cover named Endangered Earth the planet of the year was found this quote "For man has reached a point in his evolution where he has the power to affect, for better or worse, the present and future state of the planet." [23]

Unlike the flourishing environmental movement in the early 1970s, at this time environmental consciousness was deeply ingrained in American people's hearts.

China: Keep up with the Pace

China established environmental protection as a basic national policy and specified a three policy system: first is integrating prevention with control. Second is the-polluters-pay tightening up of environmental management to include payment of pollution clean-up. Third is reaping benefit from environmental protection.[24]

In 1978, China established the National Committee for Man and Biosphere. In 1980, China and the United States signed documents on cooperation in environmental protection, which initiated the process of China's bilateral cooperation in environmental protection. In 1983, China established the first national-level professional newspaper on environmental protection—"China Environment News". The second national environmental protection conference convened in 1983 emphasized environmental protection as a basic national policy of China. This strategic policy of environmental protection of China included: synchronized planning, implementation and development in terms of economic and urban and rural development and environmental protection, bringing about harmony of economic returns and contribution to society and environmental protection. On September 11, 1989, China signed the "Vienna Convention for the Protection of the Ozone Layer," which was enforced in China on December tenth of the same year. Then China also joined "Montreal Protocol on Substance" in 1991 and continues to raise people consensus on the protection of the Ozone Layer. The Environmental Protection Law of the People's Republic of China adopted at the 11th Meeting of the Standing Committee of the Seventh National People's Congress on December 26, 1989, and replaced the Environmental Protection Law of the People's Republic of China (for Trial Implementation). "As the basic law of environmental protection, it played a significant role."[25]

1990s

The United States: Inner Consciousness

In 1992, the United States had more than ten thousand non-governmental environmental protection organizations. Members of the largest 10 organizations increased from 500 thousand in 1965 to 7.2 million in 1990. For example, members of the Sierra Club were 114 thousand in 1979, and reached 325 thousand in 1982.[26] 1990 Gallup surveys found three-quarters (76% and 73%, respectively) of the public responding "yes" when asked, "Do you consider yourself an environmentalist?" Thus, it is not surprising that a 1990 survey by Environment Opinion Study (1990) found two-thirds of the public agreeing that "threats to the environment are as serious as the environmental groups say they are" and only one-quarter saying that "environmental groups are exaggerating these threats in order to get the public to pay attention to them."[27] So, as the development of the environmental protection movement, the public's environmental consciousness was also growing.

Environmental issues have become a topic which people can hear from the media and will talk about every day. Environmental scientists are often invited by the media to talk about science, technology and the environment. Sociologists also became providers of environmental concern information daily accessed by the public.

The environmental consciousness trend extended from concern in the United States to concern around the world. The United States government also took the opportunity to construct programs promoting the environmental point of view to other countries, and used political intervention in attempts control the pollution and even the policy of developing countries.

China: Reference and Development

Since 1990, China has paid more attention to the environmental protection work by holding the third, fourth, fifth and sixth national environment protection conferences. The Central government held the one-child policy and the environmental protection conferences in 1997 and 1998. From then on, population, resource, and environmental issues became routine topics of conferences in the NPC. The CPPCC put forward the transformation that turned concern mainly from pollution control to equally emphasizing attention on pollution control and ecological protection. In the 1990s, NPC and CPPCC put forward the transformation from pollution control to pollution control and ecological protection; from terminal treatment to source and entire control; from discrete point treatment to comprehensive treatment; from density control to combine density and total control. The conference also decide to operate clean process and circle economy and concentrated human resource, material resource and finance resource on the important environmental issues area—"three rivers" (Huai River, Hai River and Liao River), "three lakes" (Lake Tai, Lake Dian and Lake Chao), "two areas" (sulfur dioxide control area and acid-rain control area), "one city" (Peking), "one sea" (Bohai Sea).[28]

In addition, China cooperated closely with international environmental protection. In 1990, China attended the "Basel Convention on the Control of Transboundary Movements of Hazardous Wastes and Their Disposal." In 1992, China attended "Convention on Biological Diversity." In 1994, China attended "United Nations Convention to Combat Desertification." In 1996, China attended "Convention on Nuclear Safety." In 1998, China attended "United Nations Framework Convention on Climate Change" (e.g., Kyoto Protocol), etc. Since the "eleventh five-year plan." the central committee promoted a requirement of accelerating the achievement of the "three historic transformation." This transformation changed policy from emphasizing economic growth but ignoring environmental protection to emphasizing both environmental protection and economic growth, with environmental protection lagging behind economic growth, to finally synchronizing environmental protection and economic growth. Moreover, this transformation changing from mainly employing administrative measures in environmental protection to comprehensive use of legal, economic, technical and necessary administrative measures to solve environmental problems. This requirement established working idea: "all-round development with breakthroughs in key areas," put forward the subject that solves environmental issues from the national macro-strategic level, formulate environmental economic policies from entire process of reproduction, let overwhelmed rivers and lakes recuperated, and promote higher integration of environmental and economic policies.[29] During the Last twenty years of the 20th century, energy consumption per unit GDP was down by 34 percent, greater than 1 percent over the last same period. China achieved its goal of triple the GDP by double the energy consumption.[30]

21st CENTURY

The United States: Reflections on Modernity

"If the forecasting of the limits to growth theorists was full of problems, it nonetheless highlighted the truism—conveniently ignored by capitalism and its economist—that infinite expansion with in a finite environment was a contradiction in terms. It thus posed a potential catastrophic conflict between global capitalism and the global environment."[31] Specifically, while the academic community of the United States re-examined the development of environmental consciousness growth into a main social concern, the recognition of challenge from environment sociology is also the symbol of environmental consciousness transformation.

The political implications made environmental policy changes during the new century. In the Clinton administration, environment policy obtained great support and developed both domestically and abroad. After George W. Bush became president, the environment policy experienced decreased attention. Shortly after he became president, he set aside the pervious environmental laws, refusing to undertake the liability protection. The current President Obama differs greatly with Bush in policies concerning the environment and climate change. On January 26, President Obama signed two related presidential memorandums. In what he called "a down payment on a broader and sustained effort to reduce our dependence on foreign oil," Obama directed the Department of Transportation to establish higher fuel efficiency standards for carmakers' 2011 model year.[32]

The global economic crisis outbreak at the end of 2008, which had its source in the United States, is inextricably linked with the environment policy of President Obama. People believe that the reorganization of the energy industry will be a reorganization of the United States' global strategic force. It will bring the United States out of the current economic, environmental and national security crisis. This energy revolution will let Obama successfully promote a new economic revolution primarily of green economy. Moreover, the establishment of clean energy structure will completely change the produce and life style of the Unite States. The energy revolution will base on the American Recovery and Reinvestment Act.[33] Energy structure revolution will be the greatest economic, social and environmental revolution in the 21st century.

China: Mobilization of the Nation

Resource-saving is the urgent task of China's development, and also is a common sustainable development demand by countries around the world. Western developed countries accumulated a lot of experience in reducing resources waste and improving the efficiency of resource use. The experience learned from the west combined with China's own national characteristics, leads China to walk in its own harmonious environment way.

Based on the thinking of previous economic growth methods and developmental paths, building a conservation and environment-friendly society and realizing sustainable development is the people's right choice to alleviate contradiction between population, resources and environment. At present, China has a serious environmental situation. With the industrial and urban development, the environment pressure continuously increases. Therefore, improving the quality of the environment is a long-term and arduous process. To address this issue, the Chinese government put forward the concept of building environment-friendly society.[34] In order to accelerate building the environment-friendly society, the Ninth National People's Congress Standing Committee 30th meeting accessed "Law of the People's Republic of China on the Environmental Impact Assessment" on October 28, 2002. The State Council issued a decision that implement the scientific development view and enhanced environmental protection in December 3, 2005. "Creating a resource-saving and environment-friendly society" development strategy in essence is a value that concern people's livelihood and social development.

In 2007, the Organization for Economic Co-operation and Development (OECD) published *OECD Environmental Performance Reviews: China*. In this review, it believes: That the "efforts (regulatory, economic instruments and policy) have not been sufficient to keep pace with the environmental pressures and challenges generated by the very rapid growth of China's developing economy nor to capture the potential economic benefits to be obtained from improved pollution abatement and nature protection. Overall, environmental efforts have lacked effectiveness and efficiency, largely as a result of an implementation gap. The weaknesses in the present system are demonstrated by the failure to achieve some of the key objectives of the 10th FYP, and the severity of environmental problems in many parts of China."[35] In addition, the Chinese Public Environmental Protection Index by the China Environment Culture Promotion Association (CECPA) is considering as the Chinese public environmental protection awareness and behavior's "barometer" and it is also the first environmental protection annual index. In 2008, the environmental pollution problem ranked third in "public most concerned social hot topics." It is the third time environmental pollution problem entered the top three rankings. Secretary General of CECPA Shaomin Zhang said: "Environmental pollution problems closely follow the 'price issue' and 'food security,' the public concern rate is 37.7 percent. Environmental pollution and various environmental events still exist as threats on the socio-economic sustainable development and personal life of public. Some of them are very serious. Environmental problem has risen to the height and level which impact of social harmony and stability." Facing the serious environmental situation, China's government did not give up efforts, but instead used the 2008 Olympic Games as an opportunity to increase environmental protection efforts and enhance environmental consciousness.

"Green Olympics" was an important vow taken by Beijing to International Olympic Committee. The so-called "Green Olympics" was comprehensive movement the goals to enhance the city's sustainable development capacity, provide a clean and beautiful environment for the Olympic Games 2008, organize the Olympic Games with sustainable development perception that protect the environment, conserve resources and protect ecological balance, minimize the Olympic Games' negative impacts to the environment, to play an exemplary role for environmental protection work all over the country, carry out environmental education activities extensively to enhance public environmental consciousness, emphasis on harmony purposes between man and nature, and to comprehensive enhance public ecological civilization literacy. [36]

In fact, the Olympic Games were like a "catalyst." With the policy, funding and public support, environmental protection concepts had been fully released. In addition to policy change, individuals all changed their behaviors to protect the environment. At the same time, the Olympics provided an opportunity to apply the laboratory phase environmental protection technology and environmental protection products. For example, in the process of building the Olympic venues, many new technologies were utilized including: storm water utilization, solar photovoltaic, the light tube system engineering, ecologic water system, recycled water utilization, high efficient energy saving lamps, water saving toilets, new energy utilization, solid waste collection and treatment. This collection of technologies of environmental protection used in the Olympic Center's facilities can be seen as a "Fair" for environmental protection design. There is no doubt that Chinese environmental consciousness and environmental management were significantly increased by the Olympic Games. Environmental policies and systems were becoming better and approaching perfection day by day. Although it is difficult to pass on environmentalism as a heritage, the Chinese will try their best to do so. Director of Beijing Municipal Environmental Protection Bureau Shaozhong Du said that, "the reduce pollution measures to will be certain as the fixed form and even more stringent after the Olympic Games. Beijing, and other cities, will play an exemplary role and have the durable influence in the coming day."[37]

COMPARISON AND DISCUSSION

Through diachronic comparison, we can see awareness of environmental issues rose rapidly in the United States from 1960s to early 1970s. In that decade "primary attention was given to air and water pollution and the importance of protecting areas of natural beauty and recreational value."[38] During the period from 1973-1974, energy crises made The Limits to Growth a guiding principle. Chemical waste made sociological attention turned to local environmental hazards and introduced the different functions of environmental protection to the public awareness. However, rain forest destruction and the decline of biological diversity turned the public attention back to the level of global level. Furthermore, global warming increasingly made environmental concern more intensive. According to the United States, the development of environmental consciousness could be concluded by Table 6.1.

Table 6.1 The Development of Environmental Consciousness in the United States

1960s	The first Earth Day on 1970s	Later 1970s	1980s	1990s	2000s
Developed rapidly	"Peaked"	Declined but better than mid-sixties	"Backlash"	"boost" and higher than early seventies	structural reflection after holding new consciousness

In Hannigan's opinion, the role of two major societal institutions—science and the mass media—play the important role in constructing environmental risks, knowledge, crises and solutions.[39] This point is verified by the U.S. environmental consciousness development process.

Professor Dunlap supported Hannigan's view that Environmental Concern came from the degradation of the environment. But *Silent Spring* brought the reflection of environmental issues to people of the United States. Mass media promoted environmental consciousness causing it to peak in the early 1970s. Then came the recession (explanation to follow); however, mass environmental consciousness had already been established. Based on this, mainstream environmental concern impelled government focus on environment protection when the policy became slack needs citation and quotations.

In China, the first reason for environmental concern is also the deterioration of the environment. "1950s wash rice and vegetable, 1960s wash clothes and irrigation, 1970s water turned bad, 1980s fish and shrimp vanished, and 1990s harmed human"—this folk song prevails over Huaihe River area which reflects water pollution history vividly. Based on those environmental issues, the government established laws and regulations for environmental protection. However, the consciousness of environment protection has not struck root in the heart of people. When situation worsening, local people have taken action to protect the environment and formed non-governmental organizations spontaneously (such as Friends of Nature). Although the influence is far below *Rome Club* which impelled the environmental protection progress of the UN, the efforts did contribute to an awakening of environmental consciousness. The author believes that Chinese environmental consciousness influenced by government environmental top-down protection system on the one hand, on the other hand, Western science, mass media and the impact of UN also took important role in raising environmental consciousness. In addition, the spontaneous discussions and actions by some NGOs have also promoted the awareness of China's environmental consciousness.

Table 6.2 The Development of Environmental Consciousness in China

1960s	The first Earth Day on 1970s	Later 1970s	1980s	1990s	2000s
No Consciousness	Awareness	Development	keep pace with international situation	Became important topic	Flourishing

Contrast Table 6.1 and Table 6.2, and we can see that environmental Consciousness started late in China. The awakening of environmental consciousness followed by the United States and UN, but the overall tendency of environmental consciousness is increasing and close to maturity.

Economic Factors' Impact on Environment Consciousness

One to mirror the environmental consciousness in the United States is the satisfaction of the government's environmental measures and investments. While the government's decisions depend on economic factors, results can be seen from the comparison. Table 6.3 gives the history of the United States economic situation since the 1950s.

The development of economic factors as well as industrial development in the United States led to the 1970s environmental movement. This movement was based on the effects of increasingly serious environmental problems and the economic strength from accumulated in the 1960s in the United States. Both of these factors increased attention to environmental issues and the expanded capacity for environmental movement. Late in the 1960s and early 1970s, the pillar industry transition and environmental degradation changed the consideration of the environment-economic cost-effective view by the public. And, it also promoted the environmental movement.

Table 6.3 History of the United States Economy

Time	Characteristic	Reason
1950-1960	continued development and prosperity of the western and southern of the United States	The United States is the most wealthy and powerful country of the world. Occupied international markets; promoted science and technology; developed industries and military; improved the level of people living.
1970-1980	decline economic status because of the crisis on 1970s,the economic development be limited	Development took a serious blow by the economic crisis. Inflation and debt became worse.
1990	Sustainable and stable development in 1990s; economy entered to a new era	socio-economic reformation; education reinforced; science and technology promotion; information industry development; traditional industries transformation

Source: *The Economic of the U.S.A.* [40]

From Table 6.3, we can see that there is a strong relationship between economy and environmental consciousness. Although there is a delayed effect, the period of booming environmental consciousness in the United States is correlated with a prosperous economy. Environmental consciousness can contribute to solve environmental problems. This perspective is similar with the upward Kuznets curve. The Kuznets curve displayed the relationship between economic development and environmental quality as inverted U-curve. Which means when economy start developing, environmental degradation worsened; but, when economic developed reached a certain stage, pollution will be controlled, and the environment quality will gradually be improved.

In addition, the event of 9.11 has had a profound impact on the United States. People began to concentrate on the protection of national security. Bush opposed the Kyoto treaty because of the strain he believed the treaty would put on the economy; he emphasized the uncertainties which he believed were present in the scientific evidence.[41]

Political Factors' Impact on Environment Consciousness

As we saw the economy factors have so many important influences on Environmental Consciousness. We also can see the role that political has played in the development of environmental consciousness. From this angle, political influence is mainly manifested in the policy aspect. The policy 42 USC Section 2151p-1, adopted by Congress through P.L. 87-195 Dec. 30, 1969 (National Environmental Policy Act, 42 USC Ch. 55) embarked environmental protection on a legal track. This policy also established the consciousness of environmental protection in the legal system. The major impact on the environment from politics is the laws and regulations. The laws shared similarities in both China and the United States. The examples are the establishment of ecological protection area, restriction on the production of pollution, and so on. In addition, politics is also associated with economy and media.

Environmental policies strongly implicated economy as the "Kyoto Protocol" that was previously mentioned. In addition, industry promotion changed by economic policies in the United States, such as taxes adjustment, subsidies increase and so on. Compared with macro-economic background, micro-economic regulation will give more specific control of individuals and businesses. Instead of building an "Economic-Technological development area" as China has done, the government of the United States will use economic policy to lead enterprises.

Policies also have very important influence regarding construction and the impetus of environmental consciousness in China. Chinese environmental protection regulations have continuously improved in recent years, as has her influence of environmental consciousness by way of publicity and policy implementation. In other words, execution of environmental protection in China was passive at the beginning stage. Because environmental consciousness had not been internalized, people's behavior needs to control by law and stimulate by micro-economic instruments.

In addition, from September 1st, 2003 the China Environmental Impact Assessment (EIA) law has become the method and system of analysis, forecasting and assessment for the project's environmental impacts. Furthermore, the mass media have also had important influence on the environmental consciousness of China. The 2008 Olympic Games is an obvious example. To improve the environmental consciousness and promote good performances during Olympic Game time, the Chinese government started "China Environmental Awareness Program" in June 2006. One of the reports was "2007 China General Public Environmental Survey" which shows that the mass media has become the most important knowledge and information channel for gaining the Environmental Protection, as seen by the fact that 81.1 percent of people accept the environmental knowledge and information through television and radio. People thought the environmental issues have become serious social problems in China. In the top 13 ranked issues, environmental pollution was addressed as the fourth, closely after medical needs, employment and the income gap.[42]

The influence from politics is not unilateral. The environmental movement in the United States gave us a model of a movement that became a channel for the grass roots aspirations and which in turn was present in presidential campaigns of the United States. Al Gore, Bill Clinton's 1992 campaign partner and the author of *Earth in the Balance* has shown the environmental desire of people. *Earth in the Balance* is the 1992 best-selling book which expressed politician willing for environmental concern and determination to protect the earth. As soon as Clinton entered the White House, he gave an "Earth Day" speech which not only shown his support of environmental protection but also inspired enthusiasm of the "Earth Day".

Environmental Movement's Impact on Environment Consciousness

Environmental organizations, environmental movement gave environmental consciousness the strong push. In the United States, the organizational and ideological roots of contemporary environmentalism are commonly traced to the progressive conservation movement that emerged in the late nineteenth century.

Dunlap and Mertig (1992) pointed that there are three waves of environmental movements during the developing time.[43] The first and second wave of environmental movement were led to legislation establishing early national parks and agencies (the U.S. Forest Service); spawned conservation organizations as Sierra Club; and emphasized the mitigation of resource problems as well as the development of resources (e.g., flood control and soil conservation) as well as the development of resources (e.g., energy through the Tennessee Valley Authority) to stimulate economic recovery. In June 1966, Sierra Club published full-page newspaper ads in the *New York Times* and *Washington Post* against building a dam that would flood the Grand Canyon.

After World War II, the third wave resulting "wilderness movement" [44] was accompanied by continued concern about the future availability of natural resources as well as by growing concerns about overpopulation and air and water quality.[45] In 1976, zoologist David Etnier, who discovered the snail darter, joined Platner, Hill and others in filing a lawsuit to stop construction of the dam.[46]

Environmental protection organizations in China have three major phases. From 1978 to the early 1990s, the first birth stage. Environmental Society Science of China set up on May 1978, which is the first environmental non-governmental organization initiated by government.

From 1990 to the beginning of this century were the development stage, environmental protection extended to the community and grass roots. A number of environmental NGOs got together and acted to achieve environmental and economic development goals. On March 11, 1994, the organization named "Friends of Nature (FON)" was set up in Beijing. It is the first environmental protect organization enrolled in the civil affair government department and later became one of the famous environmental NGO in China.[47]

In 1995, FON initiated the protection of Yunnan snub-nosed monkey and the Tibetan antelope, which is the first peak of environmental NGO in China. They launched a series of publicity activities from the public protection concerned and set a good public image during this period. In 1999, another NGO named "Global Village of Beijing" worked with Beijing Municipal Government practice "Green Community" extends the grass-roots environmental protection work to public.

In 2003, the "River hydropower disputes" marked as big event in environmental movement for it made dam project given place to environment. Premier Wen Jiabao said "this kind of large project which paid highly attention of society and opposed by environmental protection, should have prudently and scientific policy-making".[48] After the early 2000s, the environmental movement entered to the mature stage with promoting public participation, giving advices and suggestions to policy, and carrying out the social supervision, and so on.

United Nations Impact

In the previous text we can see the awakening of the Western environmental consciousness and environmental movement (especially the United States) estimated by the concern of United Nations. At the same time, the UN concern strengthened environmental consciousness all over the world. The UN reinforced environmental consciousness with by giving feedback to the United States and awakened Chinese. Then, although decisions always depend on every situation, policy, environmental movement and United Nations joined as environmental allies.

On June 16th 1972, the Conference on the Human Environment at Stockholm was the UN's first major conference on international environmental issues, and marked a turning point in the development of international environmental politics. In 1977, the United Nations Environment Program (UNEP) proposed the concept "eco-development". In 1980, UNEP convened a seminar to discuss the relationship between population, resources, environment and development. The seminar pointed out that these four factors closely related to each other and inter-promoted to each other. Therefore, the management of the relationship between these four elements is the new development strategy. In 1981, Lester Brown, who gave the first system conception of sustainable development with the book Building a Sustainable Society analyzed a series of economic development issues, such as land desertification, resource depletion, oil shortages, lack of food; and put forward the channels to control population, protect resource, and exploit renewable resources. In February 1982, Our Common Future put forward the "sustainable development" concept and founded the fundamental approaches for sustainable development. On May 1982, an extraordinary session of the UNEP introduced the "new environment" concept. In 1983, the World Commission on Environment and Development (WCED) was created to address growing concern "about the accelerating deterioration of the human environment and natural resources and the consequences of that deterioration for economic and social development."

In 1992, the Union of Concerned Scientists (UCS) wrote "World Scientists' Warning to Humanity" with a majority of the Nobel Prize laureates signatures. At the same year, two programmatic documents — "Rio Declaration on Environment and Development" and "Agenda 21"—published on the United Nations Conference on Environment and Development. Realized environmental issues are the major problems faced by human beings and have a close relationship with development of society, the research of sustainable development has extended from academic to international focus and governments' agenda.

At the end of 1997, more than 150 countries and regions gathered in Kyoto, Japan in order to save the earth from the threat of global warming, by holding the United Nations Framework Convention on Climate Change (UNFCCC or FCCC). The international environmental treaty produced at the United Nations Conference was intended to achieve "stabilization of greenhouse gas concentrations in the atmosphere at a level that would prevent dangerous anthropogenic interference with the climate system." So far in 1999, including China, already 40 countries have joined the "Kyoto Protocol"; as of January 2009, 183 parties have ratified the protocol, which entered into force on 16 February 2005.

As long as recognized Environmental Sociology is the academic subject of seeking the relationship between environment and society while Politics is another academic subject of excavating the society by power and authority, it is not difficult to understand the link between politics and ecology. There is no political system which has not influenced on ecological processes.

CONCLUSION

To improve public scientific literacy is the necessary requirement of modern societies' civilization. This opinion became the consensus of both the scientific and political arenas. In our daily life, the success of a series of social development issues mostly related to the environment—such as global warming, human cloning, AIDS prevention, and the investment on technology research and so on—are mainly based on the consciousness and participation of people. Therefore, research concerning the establishment of environmental consciousness has helped us in dealing with environmental problems.

Hannigan pointed out that "Environmental problems do not materialize by themselves; rather, they must be 'constructed' by individuals or organizations who define pollution or some other objective condition as worrisome and seek to do something about it".[49] This view is related to the mass media "agenda-setting" theory. A well-known dictum by political scientist Bernard Cohen (1963) was that, "The press may not be successful much of the time in telling people what to think, but it is stunningly successful in telling its readers what to think about".[50] In the United States, important issues are put on agenda by mass media; the typical example is environmental protection in the 1970s. Agenda-setting theory shared the same idea with Anthony Downs's "issue-attention cycle"—the rise and fall of an issue on the public agenda.[51] However, Riley E. Dunlap pointed out that "Downs' issue-attention cycle is not an inexorable process".[52] Public support for environmental protection has gone through its infancy, arousal, climax, and decline, but is still in sight. Even after the government withdrew its attention from the environment, mass environmental consciousness made the 1980s revive after the recession.

The government indeed received pressure from the environmental movement. On one hand, the environmental movements made the government execute a more real practice while on the other hand, the Green Party grew and gained the opportunity to exercise influence over the government. In order to gain the public's trust and votes, American political parties called themselves environmental protectors.

The Chinese government took responsibility of the society as environment deterioration became more serious. The people related their expectations of environmental protection to the government while the government issued top-down policy for dealing with environmental issues. In China, environmental protection reflects public reliance on government. It also reflects the government's concern and responsibility for the society. The establishment of public environmental consciousness relied on science and technological development as well as the policy publicity through mass media. In addition, particular programs (i.e., the 2008 Beijing Olympic Games) were also used in an agenda-setting manner to raise the level of both environmental consciousness and the action of environmental protection.

The scientific and technological revolution opened the way for modern society, but they also created the possibility of risk to society. For developing countries, it is difficult to cross this gap. They should face the problems that developed countries have experienced, such as rule of law incompletion, economic market imperfection, rapid population grown, labor confliction, sexual discrimination, wealth differentiation, environmental pollution and so on. At the same time, developing countries faced the problems were not faced by developed country, such as nationalism and cultural conservatism. In addition, various developed levels in one developing society made it difficult to adapt to change. Furthermore, with the era of globalization, developing countries also have to face the open possibility of risk from developed countries.

Environmental problems are the typical issues include in these topics. Along with the strengthening of globalization, environmental problems have the potential to polymerize or cover the traditional problems; even occupying the position of a major problem. In this way, environmentalism will become more complex and extensive in developing countries.[53]

In regards to China's search for solutions to its environmental problems, the experience of Western society is worth learning from. However, China should not give up another road—the traditional culture of life in harmony between nature and humans. Lao Zi, Zhuang Zi, Confucius and other well-known sages often put "harmony" as the core of though. The Moral sutra said: "Ren Fa Di, Di Fa Tian, Tian Fa Dao, Dao Fa Zi Ran" means all things have to follow the nature law. Spring-autumn of Lü states, "Why not drain water for fish? No fish next year" these traditional verses show how to use natural resources rationally.

From the late 19th century to mid-20th century, Chinese lost the numerous outstanding traditional cultures and the principles of moral excellence described by its ancestors. As people are feeling more and more unsafe today, new direction of China may not be the same as the West—following the footsteps of purely technology. But rather, China may find the implicit clues from its own the history.

NOTES

* The authors want to thank Dunlap, E. Riley, the Regents Professor in the Department of Sociology at Oklahoma State University, for his valuable advice in writing this article.

1. Liu, Shaokang. 2002. *Introduction of Environmental Protection*. Qinghua University Press.
2. Dunlap, Riley E. and Jones, Robert E. 2002. "Environmental Concern: Conceptual and Measurement Issues." *Handbook of Environmental Sociology*. Edited by Dunlap, Riley E. and Michelson, William. Westport, CT: Greenwood Press.
3 .. Hannigan, John A. 1995. *Environmental Sociology: A Social Construction Perspective*. London: Routledge.
4. Catton, William, Jr. and Dunlap, Riley E. 1978. "Environmental sociology: a new paradigm," The American Sociologist, Vol. 13 pp.41-9.
5. Hannigan, John A. 1995. *Environmental Sociology: A Social Construction Perspective*. London: Routledge.
6. Eighth plenary meeting of the 8th China National Congress. August 16, 1959. *The resolution of increase production and conserve resources*. News of the Communist Party of China.
7. Hong, Dayong. 1999. "Western Environmental Sociology," *Sociology*, Vol.2
8. Hong, Dayong. 2000. Social Transition and Environmental Issues in Contemporary China—A Preliminary Analytical Framework. *Southeast Academic Research*, Vol.5
9. Hong, Dayong. 2006. Measurement of Environmental Concern: Application of the NEP Scale in China. *Society*. Vol.5
10. Chen, Ajiang. 2007. From Outer-originated Pollution to Inner-originated Pollution—Social-cultural Logic of Water Pollution of Tai Lake, *Academia Bimstris*, Vol.1
11. Carson, Rachel. 1962. *Silent Spring*. Boston: Houghton Mifflin.
12. Qu,Geping. 2000. *Dream and look forward: Past and Future of Environmental Protection in China*. China Environmental Science Press.
13. Wu, Ze. 1950. *Geographical Environment and Social Development*. Shanghai Tangli Press.
14. Wang, Zhengde. 1955. *The relationship of geographical environment, population and social development*. Shanghai: New Knowledge Press.
15. Dunlap, Riley E. and Scarce, Rik. 1991. The polls—poll Trends Environmental Problems and Protection, *Public Opinion Quarterly*. Vol. 55: 651-672.
16. Mitchell. R. C. and Davies. J. C. 1978. *The United States Environmental Movement and Its Political Context: An Overview*. Washington D.C.: Resources for the Future.
17. Dunlap, Riley E. "Trends in public opinion toward environmental issues: 1965-1990." Pp. 89 in *American environmentalism: The U.S. Environmental Movement, 1790-1990*. Edited by Dunlap, Riley E. and Mertig, Angela G. 1992.
18. Russell E. Train. 1996. The Environmental Record of the Nixon Administration. *Presidential Studies Quarterly*, Vol. 26, No. 1, The Nixon Presidency (Winter, 1996), pp. 185-196
19. Meadows, D. H., Meadows, D. L., Randers, J. and Behrens, W. W. 1972. *The Limits to Growth*. Universe Book, New York.
20. Ekins, Paul. 1993. "'Limits to growth' and 'sustainable development': grappling with ecological realities." *Ecological Economics*, Vol. 8, Issue 3, Pp. 269-288.
21. Dunlap, Riley E. 1995. "Public Opinion and Environmental Policy." Page 63-114 in J. P. Lester (ed.), In *Environmental Politics and Policy*, 2nd ed. Durham, NC: Duke University Press.

22. Qu, Geping. 2002. "The path of development from Stockholm to Johannesburg." *Chinese Environmental Protection Industry. Vol.12.*
23. Time Magazine, Jan 02, 1989.
24. Zhou, Shengxian. 2009. "Scientific Concept Implementation and Environmental Protection Development." *Seeking Truth* Vol.5.
25. Citation from Ministry of Environmental Protection of the People's Republic of China
26. From Sierra Club history on website. See http://www.sierraclub.org/history/timeline.asp
27. Dunlap, Riley E. "Trends in public opinion toward environmental issues: 1965-1990." Pp. 89 in *American environmentalism: The U.S. Environmental Movement, 1790-1990*. Edited by Dunlap, Riley E. and Mertig, Angela G. 1992.
28. Zhou, Shengxian. 2009. "Scientific Concept Implementation and Environmental Protection Development." *Seeking Truth* Vol.5.
29. Zhou, Shengxian. 2009. "Scientific Concept Implementation and Environmental Protection Development." Seeking Truth Vol.5.
30. Zeng, Peiyan. 2005. Zeng Peiyan in building a conservation-oriented society the opening ceremony of symposium. See http://hzs.ndrc.gov.cn/zdhf/t20050711_33374.htm
31. Foster, John Bellamy. 2002. *Ecology Against Capitalism*, New York: Monthly Review Press.
32. See http://www.america.gov/st/env-english/2009/January/20090127161856 lcnirellep 9.743899e-02.html
33. Write House. 2009. *American Recovery and Reinvest Act*, U.S. Government Documents.
34. Hu, Jingtao. 2005. The Communist Party of China Central Committee's Proposal on the Eleventh Five-year Program on National Economy and Social Development.
35. Organization for Economic Co-operation and Development. *OECD Environmental Performance Reviews: China*, OECD Publications.
36. The official website of the BEIJING 2008 Olympic Games. See http://www.china.com.cn/sport/zhuanti/2008ay/2008-08/02/content_16118236.htm
37. Du, Shaozhong. 2008. See the news from Xinhuanet.http://news.xinhuanet.com/_olympics/2008-08/24/content_9671839.htm
38. Dunlap, Riley E. and Marshall, Brent K. 2007. "Environmental Sociology." Pp. 329-340 in Clifton D. Bryant and Dennis L. Peck (eds.) *21st Century Sociology: A Reference Handbook*. Vol.2. Thousands Oaks, CA: Sage.
39. Hannigan, John A. 1995. *Environmental Sociology: A Social Construction Perspec tive*. London: Routledge.
40. Yukun, Zhu & Xianjun, Sun, 1990. *The Economic of the U.S.A.* People's Press, Pp 249-252.
41. Barrett, Scott. 1998. "Political economy of the Kyoto protocol." *Oxford Review of Economic Policy*. Vol. 14, NO. 4. Pp. 20-39.
42. China General Public Environmental Survey. 2007. China Environmental Awareness Program result by Chinese Academy of Social Sciences. http://www.chinaceap.org/electrion.html#
43. Dunlap, Riley E. and Mertig, Angela G. "The Evolution of the U.S. Environmental Movement from 1970 to 1990: An Overview." Pp.1 in *American environmentalism: The U.S. Environmental Movement, 1790-1990*. Edited by Dunlap, Riley E. and Mertig, Angela G. 1992.
44. McCloskey, Michael. 1972. "Wilderness Movement at the Crossroads, 1945-1970." *The Pacific Historical Review*. Vol. 41, No. 3.

45. Paelke, Robert. 1989. *Environmentalism and the Future of Progressive Politics*. Yale University Press, Pp. 273-283.
46. Environmental Movement Timeline, A History of the American Environmental Movement, Ecology Hall of Fame. http://www.ecotopia.org/ehof/timelinetext.html
47. From FON website. See http://www.fon.org.cn/channal.php?cid=2
48. From ngocn.org website. See http://news.ngocn.org/?action-viewnews-itemid-909
49. Hannigan, John A. 1995. *Environmental Sociology: A Social Construction Perspe ctive*. London: Routledge.
50. Cohen, Bernard Cecil. 1963. *The Press and Foreign Policy*, Princeton University Press.
51. Downs, A. 1972. "Up and Down with Ecology—the 'Issue-Attention Cycle.'" *Public Interest* 28:38-50.
52. Dunlap, Riley E. 1995. Public Opinion and Environmental Policy. Page 63-114 in J. P. Lester (ed.) Environmental Politics and Policy 2nd ed. Durham, NC: Duke University Press.
53. Yan, Min. 2007. "Green Opinion Rise in China." *Sociological Perspective*, Vol.11

Chapter 7

Urban Environmental Protection Policies and Their Effects in China

Li Yujun

THE PROCESS OF URBAN ENVIRONMENTAL PROTECTION POLICY-MAKING IN CHINA

According to the guidelines of environmental protection and the transformation of management, there are three phases of urban environmental protection in China. 1973 to 1983 was the initial starting phase of the environmental protection, especially focusing on point source treatment of industry and comprehensive utilization; 1984 to 1996 was the improvement phase, during which the existing law and regulation system of environmental protection shifted from single pollution control to integrated control; from 1996 to the present has been the deepening and innovating phrase of environmental protection.[1]

Initial Starting Phase of Environmental Protection (1973-1983): Point Source Treatment of Industry and Comprehensive Utilization

In 1972, a several serious environmental events happened in China. Pollution in Dalian bay was emergent, the ebb and flow of the sea was turbid, every year the loss of sea cucumber was more than 10,000kg, the loss of shellfish is over 100,000kg, the loss of corbicula is no less than 1,500,000kg; the water pollution of Beijing Guanting reservoir precipitated a massive surfacing of dead fish; the pollution of Songhua river's water system also raised alarm. In June 1972, under the leadership of Premier Zhou Enlai, a Chinese government delegation attended the conference on human development held by the United Nations in Stockholm Sweden. After this conference, the Chinese delegation became deeply convinced that environmental issues have great influence on economic and social development, and senior policy makers recognized that there are serious environmental issues in China deserving of serious treatment.[2]

In August 1973, the first Chinese environmental protection conference was convened, and the conference issued "thirty two characters" environmental protection guidelines and the first file of environmental protection in China named "several regulations of environmental protection and improvement." On 17 November 1973, in order to efficiently control industrial pollution, the National Planning Commission, the National Construction Commission, and the Ministry of Health promulgated the first environmental standards in China named "Trial emission standards for three wastes from industry." In February 1978, the First Session of the Fifth National People's Congress approved the Constitution of the People's Republic of China. The new Constitution prescribed that "The State protects and improves the environment in which people live and the ecological environment. It prevents and controls pollution and other public hazards." The Chinese government attaches great importance to environmental protection, clearly putting forward that environmental protection is an important part of building socialist modernization. In September 1979, the eleventh Standing Committee of National People's Congress of the Fifth National People's Congress approved the first environmental protection basic law in China named "Environmental Protection Law of the People's Republic of China (trial)," establishing three regulations, including environmental impact assessment, the regime of the "three-at-the-same-time" system (i.e., facilities for preventing and controlling environmental pollution and destruction shall be planned, constructed and put into use at the same time as the main production projects) and pollution charges. Chinese environmental protection measures were thereby brought into the legal regime.

During the ten years from 1973 to 1983, industrial pollution became relatively serious, thus the major thrust of environmental protection focused on industrial point source treatment and the comprehensive utilization at the initial starting process of the environmental protection.

During the initial starting process, government and academic circles had yet to develop clear theoretical approaches to environmental protection; for instance, how to deal with relationship with the economic construction, social development and environmental protection? How to develop Chinese environmental protection? Do economic policies have great influence on environment? Meanwhile the problems in environmental protection practice were also still vague; such as, how to control industrial pollutions? Who should supervise and manage environmental protection works of enterprises and institutions? How can we turn harm into utilization to achieve comprehensive utilization? Although many theoretical and practical problems remain, by means of continuous trial and experimentation, the Chinese government has transformed the cognition of environmental issues gradually, recognizing that environmental issues are not simply "three wastes" problems but are also closely connected to economic development. Moreover, the Chinese environmental protection policies, laws, regulations and standards had been established, pursuing legalization of environmental protection. The Chinese government had also established national and provincial management organizations, and therefore the basic framework of the Chinese environmental management regulation system had taken shape.

Consummated Phases of Environmental Protection (1984-1996): Integrated Control of Pollution

A second Chinese environmental protection conference in December 1983 clearly stated that environmental protection was a basic national policy. To effectively prevent and control industrial pollution, the government drew up three major policies for environmental protection, i.e., "putting prevention first and combining prevention with control," "making the creator of pollution responsible for treating it" and "intensifying environmental management."

The third environmental protection conference in April 1989 proposed a system of objectives and responsibility of environmental protection. In addition, it established the policy on the quantitative evaluation of overall improvement of urban environmental conditions, the policy on pollution emission licensing, and the policy on the centralized control of pollution and the time-limit control management. It was made of "eight policies on environmental protection."

In establishing a guiding concept for the prevention and control of industrial pollution, the second Chinese industrial pollution management working conference in October 1993 decided on "three changes," i.e., (1) regarding basic strategy, China will gradually change its strategy of end-of-pipe pollution control into pollution control during the whole process of industrial production; (2) with respect to the control of pollutant discharge, concentration control will be replaced by a combination of the control of concentration and that of total quantity; and (3) with regard to pollution control methods, the focus on the control of scattered point sources will be replaced by a combination of centralized and scattered controls. [3]

During this period, the government formulated, revised and promulgated several environmental protection laws, regulations, standards and management ordinances, had issued a series of industry, trade and technology policies, economic and technological regulations for the benefit of environmental protection. Furthermore, the government formulated strategies and measurements relating to international guarantees. Compared to the initial starting phrases, every law, regulation and policy had been improved and become more operable during this period of time, and environmental protection work had achieved great progress. This progress is represented mainly in five aspects: first of all, environmental protection was established as a national basic policy, bringing environmental protection planning into the national economic and social developmental program. Economic development was established as the premise of environmental protection, striving to keep economic construction, social development and environmental protection in balance. Second, in legal provisions clearly stated the responsibilities of local governments, environmental protection management departments and institutions. Third, the institutional system of environmental protection regulation and management had been improved continuously. Fourth, the organizational construction of environmental protection management had been strengthened, building national, provincial, city level and county level independent environmental protection management institutions as powerful guarantees for the practice of environmental policy. Fifth, as there were "three changes" in the theory of national industrial pollution management, management work achieved significant progress. There were great deals of environmental issues to address during this period, bringing a great deal of pressure on the Chinese government.

The Deepening and Innovating Phase of Environmental Protection (1996-present): Implementing Scientific Development View and Constructing an Environmental Friendly Society

The fourth Chinese environmental protection conference in July 1996 issued "Several Decisions on Environmental Protection Issues." The conference stated that the implementation of environmental protection is a key part of the sustainable development strategy, and to preserve the environment is to preserve productivity. In order to achieve the goal of environmental protection, "The National Plan for Total Amount Control of Major Pollutants Discharge in China" and "China Cross-Century Green Engineering Program" had been compiled and brought into being, and the "33211 projects" had been implemented, including three rivers, three lakes, two districts, one city and one sea. The core of the management strategy had been transformed from pollution treatment to protection and control together with ecological protection.

The fifth Chinese environmental protection conference on 8 January 2002 proposed that environmental protection is one of the important functions of government, entreating the government to mobilize the whole society to join in this job in accordance with the socialist market ecology demand.

The National Council issued "Decisions on Implementation of Scientific Development View and Reinforcement of Environmental Protection." In order to implement the scientific development view, accelerate the construction of a harmonious society, and accomplish the goal of comprehensive construction a well-off society, environmental protection constitutes a key strategy; economic and social development and environmental protection should be in perfect balance.

To achieve this goal, the Chinese government sped up the "three changes." First, changing from emphasizing economic growth but ignoring environmental protection to emphasizing both environmental protection and economic growth; second, changing from environmental protection lagging behind economic growth to synchronizing environmental protection and economic growth; and third, changing from mainly employing administrative measures in environmental protection to comprehensive use of legal, economic, technical and necessary administrative measures to solve environmental problems. China should establish an ecological civilization and an environmental friendly society.

The goal of national pollutant emission control of the "eleventh five-year plan" is to, by the end of 2010; reduce energy consumption per unit of gross domestic product by 20 percent and major pollutants discharge by 10 percent. By the end of 2008, the country had registered a drop in chemical oxygen demand (COD), a key test of water pollution, and the total emission of sulfur dioxide (SO_2), the chief air pollutant, by 6.61 percent and 8.95 percent respectively from 2005 levels.[4]

MAJOR MEASURES OF CHINESE URBAN ENVIRONMENTAL PROTECTION

By the end of 1978, the population was 962.59 million, among which the urban population was 172.45 million, only 17.92 percent of the total; the urbanization rate was therefore lower than 18 percent. By the end of 2007, the urbanization rate reached 44.9 percent. During 30 years, Chinese urbanization levels increased constantly, entering a rapid development period. 70 percent of GDP and 80 percent of the state's tax income came from urban areas.[5]

Along with the growth of the urban population, the development of the economy, and continuous improvement of the people's consumption levels, industrial and domestic pollution caused very serious damage to the urban environment; the original urban ecological system is out of balance, raising a lot of unsolved problems of urban environment, thus, urban environmental protection had always been the foundation of national environmental protection policy.

Presently, Chinese urban environmental management mainly implements four measures, which include the objectives and responsibility system of environmental protection, report regulations of urban air quality, comprehensive management and quantitative evaluation of urban environmental conditions, and the establishment of National Model Cities of Environmental Protection (NMCEP). These four measures have efficiently promoted the development of urban environmental protection.

Objectives and Responsibility System of Environmental Protection

The objectives and responsibility system of environmental protection is in form of responsibility agreements; the agreements regulate certain goals and assignments during the tenures of provincial governors, mayors, and county governors, and it is key measure of achievement assessment.

Report Regulation of Urban Air Quality

Report regulation of urban air quality refers to local air quality index which reflects the urban air quality and is released to the public. In 1997, the urban air quality weekly report was primarily published in 13 major cities, including Beijing, Shanghai, Chongqing, Dalian, and Xiamen. From 2000, CCTV and some major newspapers published air quality daily reports which included 40 cities' air quality information. Nowadays, 113 major environmental protection cities have implemented the air quality report regulations, and some of the cities have undertaken air quality forecasting. These measures are significant for the development of public environmental consciousness, environment supervision and the improvement of urban environment conditions.

Comprehensive Management and Quantitative Evaluation of Urban Environment

On 13 July 1988, the environmental protection commission of the state council issued "Decisions on comprehensive management and quantity evaluation of urban environment," setting comprehensive environmental management as an important duty of government, requiring mayors to take responsibility for environmental quality, make it environmental quality an important part of their political achievement evaluation. According to the decision, the evaluation measures started on 1 January 1989. As one of the regulations of urban environmental management in China, the evaluation measures had been established and practiced widely in China, dramatically promoting the political importance of urban environmental protection.

Since the implementation of evaluation measures in 1989, urban environmental management transformed from being qualitative to quantitative, from being experiential to scientific. The implementation of comprehensive management of the urban environment organically integrates pollution management and urban infrastructure construction, transforming simple pollution management into adjustment of industrial structure and city patterns. In twenty years, the evaluation measures strengthened government management of environmental protection, accelerated environmental infrastructure construction and the improvement of the environment, and promoted the position of environmental protection in comprehensive urban development decisions.[6] At the same time, as a result of the evaluation measures, the professional quality of environmental protection departments of cities has been improved dramatically.

The evaluation measures include four standards: environmental quality index, pollution control index, environmental construction index and environmental management index. For the purpose of keeping pace with changes and achieving maximal efficiency of the evaluation measures, the urban ecological evaluation index had been supplemented constantly. Since 2002, Chinese urban environmental management and comprehensive treatment annual report has been published every year.

Since 1989, the Environmental Protection Agency implemented the quantitative evaluation measures in municipality cities, provincial capitals and other 32 cities, such as Dalian, Suzhou, and Guilin; by 1996, the number of evaluation cities was 47. By 2002, the number of major cities included rose to 113. By the end of 2007, there are 617 evaluated cities in total. [7]

Establishment of National Model Cities of Environmental Protection

After the United Nations Conference on Environment and Development in 1992, sustainable development had become contemporary worldwide goal. Sustainable development of cities received high attention and became a high priority, as cities are the center of the political, economic and cultural systems. For the purpose of sustainable development, the "ninth five year period plan of national environmental protection and 2010 long-term goals" claimed that several model cities should be built up which are with rapid economic development, tidy and nice environment, and fine ecological system.

The Environmental Protection Agency launched the activity of national model cities of environmental protection (NMCEP) in 1997. The activity mobilized the enthusiasm of urban governments and social sectors, promoting environmental protection. The Environmental Protection Agency (EPA) established the assessment index system of NMCEP, including basic standards and an assessment index, 27 indexes in total. The assessment index includes socio-economic, environmental quality, environmental construction and environmental management measures. In order to regulate and supervise the construction of NMCEP, refine the assessment index, and make the model cities into exemplars which promote the construction of resource-saving, environmental friendly and ecological civilized society, EPA revised and improved the assessment index in 1998, 2002 and 2008. EPA built up an expert database to investigate and assess the cities that applied for model status, and the cities which met the standards would be granted as "NMCEP." A review system would ensure the cities' continuous improvement.

THE ACHIEVEMENT OF URBAN PROTECTION AND CONTROL IN CHINA

Nowadays, Chinese urban environmental protection conditions can be summarized as generally stable, locally improved, and individually deteriorated. After a 30 years endeavor, the state of the Chinese urban environment has improved, but there are still some unsatisfactory aspects. Theoretically, urban ecological repair is a long-term process; it will take many years to realize goals, thus support from the whole society is required to promote urban environmental protection continuously to make the dream of "clear water, blue sky" become true.

Improved Urban Environmental Quality

Urban Air Quality Has Been Improved Continuously

Chinese urban air quality is improving continuously. The annual concentration of particulates and sulfur dioxide in the air had decreased. Considering different city scales, the air quality in some of the big cities has been improved dramatically because of energy structure adjustments; the air quality in some of the small cities has deteriorated, due to the rapid development of the economy, the increase of pollutant emission volume, and the lack of the environmental protection facilities. The proportion of cities with air quality at or better than Grade II standard was 36.5 percent in 2000; by the end of 2008, it increased to 71.6 percent, while the proportion of cities with air quality worst than Grade III standard was decreased from 33.1 to 1.5 percent (See Table 7.1 and Figure 7.1)

Table 7.1 Proportions of Cities with Varied Levels of Air Quality (2000-2008)

Year	2000	2001	2002	2003	2004	2005	2006	2007	2008
At or better than Grade II (%)	36.5	33.4	34.1	41.7	38.6	60.4	62.4	60.5	71.6
Grade III (%)	30.4	33.4	34.7	31.5	41.2	29.1	28.6	36.1	26.9
Worst than Grade III (%)	33.1	33.2	31.2	26.8	20.2	10.6	8.9	3.4	1.5

Source: Ministry of Environmental Protection of PRC, *Report on the State of the Environment In China, 2000-2008.*

Figure 7.1 Proportion of Cities with Varied Levels of Air Quality (2000-2008)

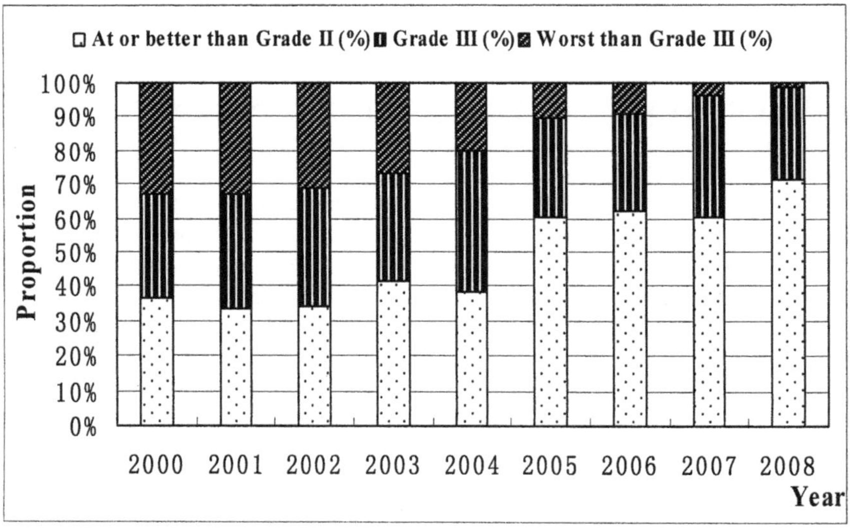

Urban Water Environmental Quality Is Relatively Stable

According to the assessment of urban water functional areas division, from 2004 to 2007, the standard rate of water quality in urban water function areas was basically maintained at 85 percent, the standards of Chinese urban drinking water is relatively high. Chinese urban water environmental quality is comparatively stable (See Table 7.2)

Table 7.2 Chinese Urban Water-Environment Quality

Year	Access standard rate of water quality	Standard access rate of drinking water	Number of the cities involved in evaluation
2004(%)	85.47	96.75	500
2005(%)	89.17	97.7	509
2006(%)[8]	82.13	88.69	595
2007(%)	86.5%	-	617

Source: Ministry of Environmental Protection of PRC, 2004-2007 *Annual Report on Urban Environmental Management and Integrated Control.*

The State of Urban Regional Acoustic Environment Is Improved Dramatically

The average equivalent sound levels of urban regional environmental noise and urban road traffic noise have improved dramatically. In 2000, the urban regional noise level of 176 monitored in cities ranged from 47.2 dB to 65.8 dB. 6 percent of the cities had a serious noise pollution; 11 percent had a good environment in terms of noise level. The district noise level in some key cities was still a little high. By 2008, the proportion of a good acoustic environment increased to 72 percent, and none of them demonstrated serious noise pollution after 2007 and 2008 (See Table 7.3 and Figure 7.2)

Table 7.3 Statistics of District Noise Levels in Cities (2000-2008)

Year	2000	2001	2002	2003	2004	2005	2006	2007	2008
Good (%)	11	11.3	3.8	54.3	62.3	60.2	68.8	72. 0	71.7
Light pollution (%)	33	33.0	48.6	42.6	34.6	33.6	29.3	26.6	27.3
Intermediate pollution (%)	49	49.4	44.2	2.5	2.9	1.7	1.6	1.4	1.0
Serious pollution (%)	6	6.3	3.4	0.6	0.3	0.9	0.3	0.0	0.0

Source: Ministry of Environmental Protection of PRC, *Report on the State of the Environment In China, 2000-2008.*

In 2000, the noise of traffic roads of 214 monitored cities ranged from 56.2 dB to 80.7 dB. 8.9 percent of the cities registered serious noise pollution. This number decreased to 0.5 percent in 2008; 15.4 percent had a good environment in terms of noise level in 2000, increasing to 92.3 percent in 2008 (See Table 7.4 and Figure 7.3)

Table 7.4 Statistical Table of Urban Road Traffic Noise (2000-2008)

Year	2000	2001	2002	2003	2004	2005	2006	2007	2008
Good (%)	15.4	25.3	13.6	79.1	84.1	86.5	88.7	92.3	92.4
Light pollution (%)	53.3	48.7	64.3	12.5	9.8	7.4	8.5	5.7	4.2
Intermediate pollution (%)	22.4	16.5	17.2	5.2	4.3	4.4	1.8	1.1	2.9
Serious pollution (%)	8.9	9.5	4.9	3.2	1.8	1.7	1.0	0.9	0.5

Source: Ministry of Environmental Protection of PRC, *Report on the State of the Environment in China, 2000-2008.*

Figure 7.2 Trend Charts of Regional Noise Levels of in Cities (2000-2008)

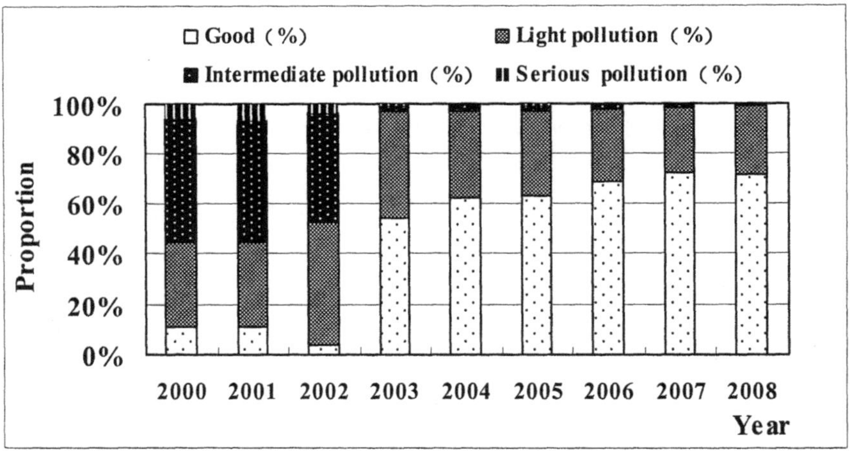

Figure 7.3 Change Trend of Urban Road Traffic Noise (2000-2008)

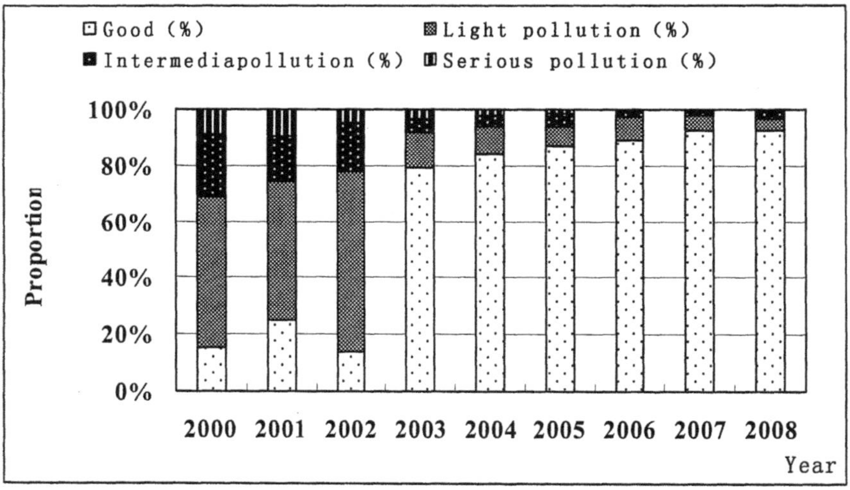

The Level of the Urban Environmental Infrastructure Is Improving Constantly

Since 1990, state and local governments have constantly increased investment in the construction of urban environmental infrastructure. The domestic waste water treatment rate, raw treatment rate of domestic refuses, urban green coverage rate, gas access rate, and the city central heat supply coverage rate have all increased.

The Investment of Environmental Protection Increased Year by Year, Promoting Environmental Infrastructure Construction

Since 1990, state and local governments have increased investment in urban environmental infrastructure, promoting the development of urban infrastructure construction. In 2000, the investment of urban environmental infrastructure was 56.13 billion RMB; in 2008, it was 180.10 billion RMB (See Table 7.5 and Figure 7.4)

Table 7.5 The Investment of Environmental Treatment (2000-2008)

Year	Total investment of pollution treatment (billion)	Investment of urban environmental infrastructure (billion)
2000	106.07	56.13
2001	110.66	59.57
2002	136.34	78.53
2003	162.723	107.2
2004	190.86	114.0
2005	238.8	128.97
2006	256.78	131.49
2007	338.76	146.78
2008	449.03	180.10

Source: Ministry of Environmental Protection of PRC, *China Environmental Statistical Yearbook 2000-2008*

The Innocent Treatment Rate of Urban Domestic Refuse Has Been Improved Gradually

Since the rapid development of economy, the acceleration of urbanization, and the improvement of public life quality in China, the amount of urban domestic refuse discharge is increasing year by year. It was about 25 million tons in 1979, and about 155 million tons in 2008 (See Table 7.6 and Figure 7.5)

The construction of treatment facilities of urban domestic refuse in China is obviously legging behind economic development; most domestic refuse treatment was simply landfill in the 1980s. The innocent treatment rate of domestic refuses was only 6.9% in 1980. After 1990, the Chinese government increased investment to deal with refuse issues. From 1980 to 1995, the rate had

Figure 7.4 The Investment of Environmental Treatment in China (2000-2008)

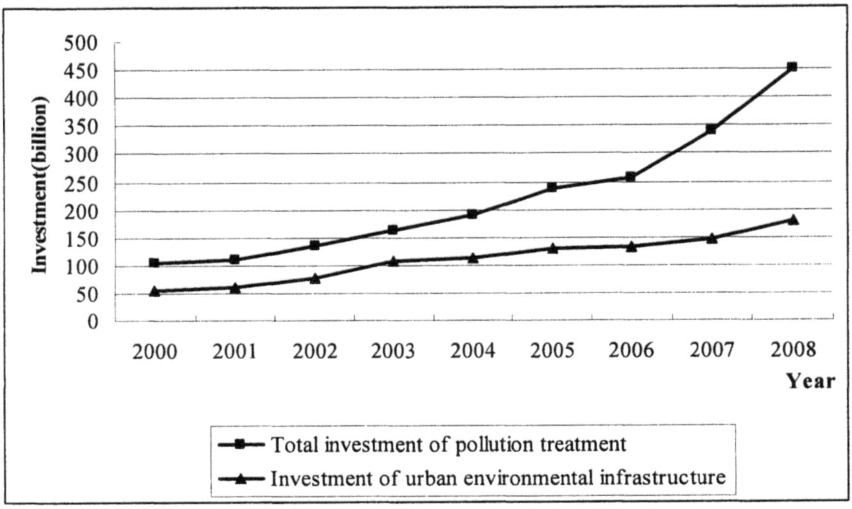

Table 7.6 The Amount of Urban Domestic Refuse Discharge in China 1979-2008 (Unit: million tons/year)

Year	1979	1980	1981	1982	1983	1984	1985	1986	1987
Amount	25	31	26	31	35	38	45	50	54
Year	1988	1989	1990	1991	1992	1993	1994	1995	1996
Amount	58	63	68	76	83	88	100	107	108
Year	1997	1998	1999	2000	2001	2002	2003	2004	2005
Amount	110	113	114	118	135	137	149	155	156
Year	2006	2007	2008						
Amount	148	152	155						

Source: The databases of 1996-2008 are original to *Urban Construction Statistical Bulletin;* The databases of 1979-1995 are original to *Management and Disposal Technologies of Municipal Solid Waste,* Administration of Pollution Control Department, MEP, China Petrochemical Press, 2000, p.10.

been improved to 56.4 percent. A lot of urban landfill sites closed during 2001-2006, and the construction of treatment facilities cannot meet the demand of domestic refuse management. The innocent treatment rate of domestic refuse has declined from 60 percent in 2001 to 52 percent in 2005 (See Table 7.7 and Figure 7.6) Urban domestic waste pollution has contributed to more and more disputes and social problems, causing every level of the city government to pay a great deal of attention to domestic waste control and management.

Figure 7.5 The Amount of Urban Refuse Discharge in China (1979-2008)

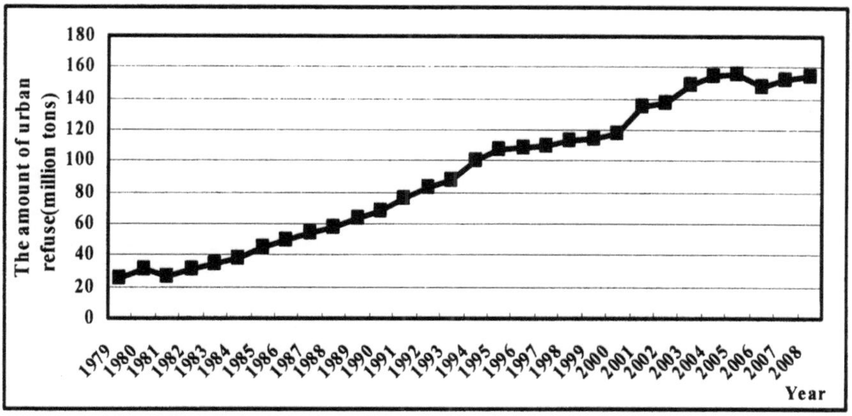

Table 7.7 The Innocent Treatment Rate of Domestic Refuse in China (1980-2008)

Year	1980	1985	1990	1995	2000	2001	2002
The innocent treatment rate (%)	6.9	5.2	3.1	56.4	58.0	60.0	54.0
Year	2003	2004	2005	2006	2007	2008	
The innocent treatment rate (%)	51.0	52.0	52.0	52.0	62.0	64.0	

Source: Ministry of Construction of PRC, *Urban Construction Statistical Bulletin 1980-2008*

Figure 7.6 The Innocent Treatment Rate of Domestic Refuses in China (1980-2008)

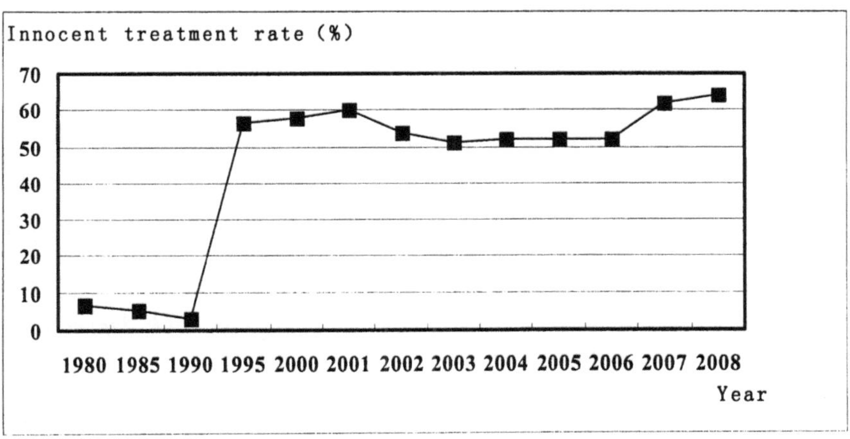

Wastewater Treatment Rate Had Improved Gradually

In 1992, the urban wastewater treatment rate was 17% in China, and it increased to 65% in 2008. The huge change is very important to the improvement of urban water environmental quality (See Table 7.8 and Figure 7.7).

Table 7.8 Statistics of Chinese Urban Wastewater Treatment Rate

Year	1992	1993	1994	1995	1996	1997	1998	1999	2000
Wastewater treatment rate （%）	17	20	18	19	23.5	26	30	32	34
Year	2001	2002	2003	2004	2005	2006	2007	2008	-
Wastewater treatment rate （%）	36	40	42	46	52	57	63	65	-

Source: Ministry of Construction of PRC, *Urban Construction Statistical Bulletin 1992-2008.*

Figure 7.7 The Urban Wastewater Treatment Rate in China (1992-2008)

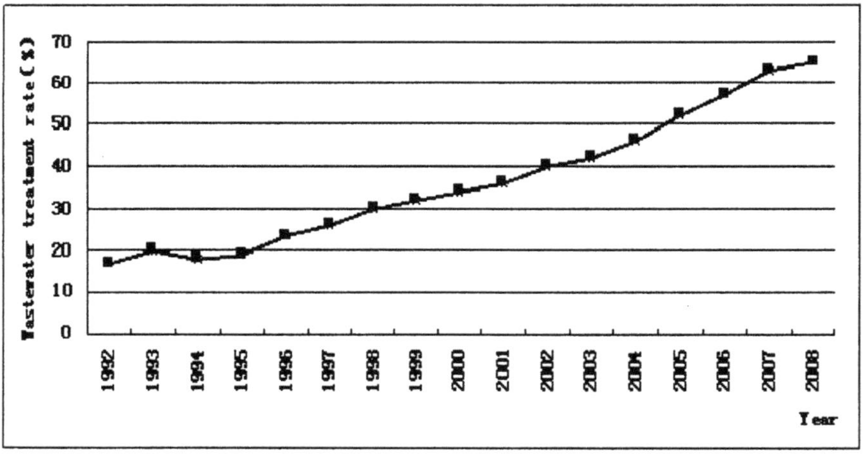

The Level of Urban Green Coverage has Improved Yearly

The green coverage rate in urban constructed areas improved from 28 percent (2001) to 35 percent (2007), urban park green space area per capita was 5.28 m^2, and it increased to 8.98 m^2 in 2007. The increased green coverage has markedly improved urban environmental quality (See Table 7.9, Figure 7.8, and Figure 7.9).

The National Model City of Environmental Protection with Harmonious Development of Economy-environment

In order to prompt the harmonious development of urban economy-society-environment, the Ministry of Environmental Protection (MEP) launched the

Table 7.9 The Park Green Space Area per Capita and Green Coverage Rate in China (2001-2007)

Year	2001	2002	2003	2004	2005	2006	2007
park green space area per capita(m^2)	5.28	5.36	6.49	7.38	7.91	7.94	8.98
green coverage rate(%)	28	30	31	32	33	34	35

Source: Ministry of Construction of PRC, *Urban Construction Statistical Bulletin 2001-2008*.

movement of national model cities of environmental protection (NMCEP) (See Table 7.10). According to statistics from the MEP, the economy of NMCEP cities increased steadily, the average growth rate increased by 53.41 percent, the growth rate of per capital GDP was 26.7 percent, the growth rate of the economy and per capita GDP of NMCEP cities are respectively 1.5 and 3.5 times of Chinese average levels, and the growth rate of environmental protection investment improved by 18.1 percent. [9]

Figure 7.8 The Park Green Space Area Per Capita in China (2001-2007)

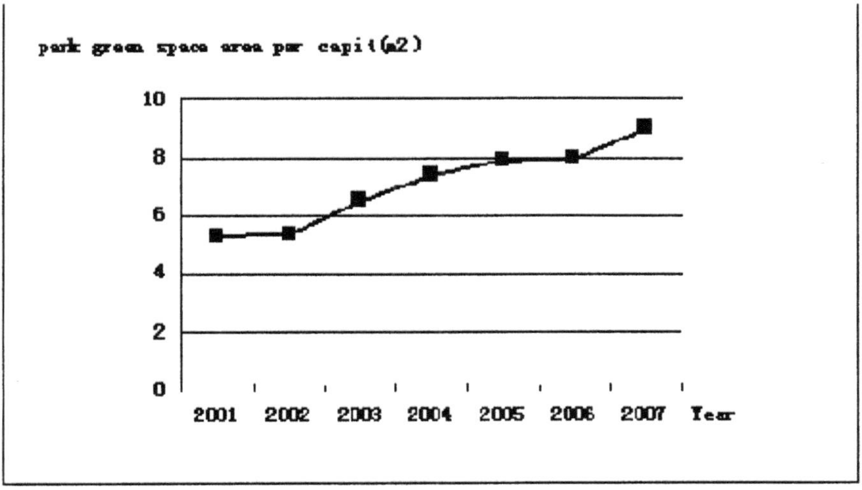

From 1997 to 2009, MEP named 67 NMCEP and 5 national model urban areas of environmental protection (See Table 7.10). Presently only 10 percent of the cities have obtained this honor, and they are mainly located in east developed part of China. How to improve urban environmental quality in the whole country is still an important research topic.

Figure 7.9 The Green Coverage Rate in Urban Constructed Areas (2001-2007)

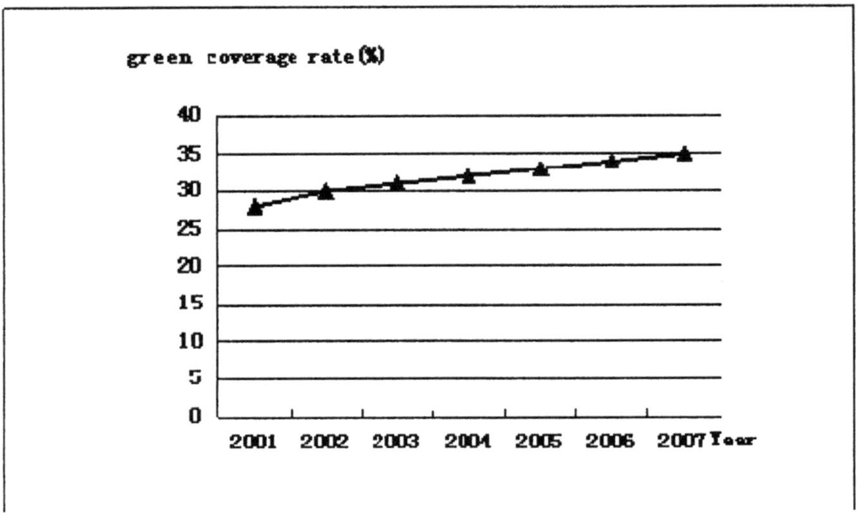

Table 7.10 List of National Model Cities of Environmental Protection

year	Number of cities	Cities named as national environmental protection model city
2007	4	Guangzhou, Shouguang, Taizhou, Yiwu.
2006	12	Tianjin, Langfang, Nantong, Huzhou, Zhaoqing, Quanzhou, Yixing, Jimo, Pingdu; Pudong District of Shanghai, Beipei District of Chongqing.
2005	9	Chengdu, Fuyang, Baoji, Guilin, Jiaonan, Laixi, Rizhao, Penglai, Weifang.
2004	12	Mianyang, Wuxi, Jintan, Liyang, Fuzhou, Zhenjiang, Chongzhou, Shenyang, Kelamayi, Kuerle, Jiangmen; Yubei District of Chongqing.
2003	3	Wujing, Nanjing, Dongying.
2002	8	Huizhou, Zhaoyuan, Shaoxing, Rushan, Haimen, Changchun, Yangzhou, Jiaozhou.
2001	4	Hangzhou, Ningbo, Changshu, Taicang.
2000	4	Qingdao, Jiangyan, Daqing, Wendeng.
1999	5	Haikou, Shantou, Suzhou; Minhang District of Shanghai; Dagang District of Tianjin.
1998	5	Kunshan, Yantai, Laizhou, Rongcheng, Zhongshan.
1997	6	Zhangjiagang, Shenzhen, Dalian, Zhuhai, Xiamen, Weihai.

Source: http://www.zhb.gov.cn/cont/mhcity/

THE CHALLENGES FACED BY URBAN ENVIRONMENTAL PROTECTION

After 30 years of endeavor by the Chinese government and people, the achievements of Chinese environmental protection are obvious, but at the same time, China's urban environmental protection faces severe challenges.

The Current Economic Growth Mode Places Increased Pressure on Urban Environments

Since 2002, many industries have developed rapidly, such as the power industry, iron and steel industry, building materials and so on. The average annual growth rate is above 10%, in which annual growth rate of iron and steel output is more than 20%. Chinese total energy consumption was 143 million tons in 2000, and it exploded to 265 million tons on 2007. The coal consumption in China is around 70% of the total urban energy consumption, and coal consumption has already become the major source of air pollution. According to estimates, 70% of the dust emission, 90% of the sulfur dioxide, 67% of the nitrogen oxides, and 90% of the carbon dioxide are caused by coal combustion in China.[10] Because of limited resources, coal will continue be the predominant part of the Chinese energy structure; it brings great challenges to the improvement of energy efficiency and the relieving of environmental pressures. Meanwhile, the Chinese urban economy maintains rapid development, and this development model causes high input, high consumption and high emission, constituting a tremendous threat to the natural environment which is the basis of urban living.

The Construction of Urban Environmental Protection Infrastructure Can Hardly Meet the Demands of Urban Development

Due to continuous urban population growth, the amount of domestic refuse shows an upward trend yearly. The environmental facilities hardly met the demands of urban development, although some of the major cities have increased the environmental investment. In 2007, among those 617 evaluated cities, 159 cities were lacking in domestic refuse raw treatment facilities, 168 cities didn't have secondary treatment of domestic water waste and emissions couldn't meet standards. Among all prefecture level and above cities, there were 38 cities without urban domestic refuse raw treatment facilities, and 46 cities without urban domestic water waste secondary treatment facilities. This lack of facilities severely influenced the improvement of urban environmental quality.[11]

The Urban Environmental Risks Are Outstanding

Recently, environmental pollution events have occurred with frequency, and the environmental risks are outstanding. Environmental risk management needs to be improved. A chemical plant blast took place on 13 November 2005 in northeast China's Jilin Province. It later spilled large amount of benzene into the Songhua River, causing a four-day water cutoff in the downstream Harbin city, capital of neighboring Heilongjiang province in late November. It was unexpected that one chemical factory's explosion accident would result in such a severe impact to the environment, economy, and society and also have international reverberations.

Meanwhile, the shortage of landfills frequently causes pollution accidents. China is presently in a peak period of environmental pollution accidents; there is one environmental pollution accident happening on average every two days. In 2007 the number of 100 environmental emergency calls received by the National Environmental Protection Agency increased by 45% from 2005. In the past ten years, the number of the environmental complaints has shot up to 710000 in 2007, and public disturbances caused by environmental issues increased 29 percent each year.[12]

Environmental issues have obviously influenced the stability of society. Thus, governments should pay much more attention to prevention of, preparation for and response to environmental pollution emergencies.

The Failure to Abide by the Law Was Common

After thirty years of exploration and implementation, China has now established a set of complete and Chinese characteristic pollution control legal systems and management regulations. Especially notable is the development of environmental legal system designation. But, there remain deficiencies in the practice and implementation. Some of the local governments, for purpose of short-term economic interests, intervene in the execution of the law, failing to implement the environmental impact assessment regulations for some new projects. They approved some high-pollution, high-consumption or irritant-distributing projects causing heavy damage to the environment, and even the treatment speed couldn't catch up with the damage speed. Thus, the failure to abide by the law influenced the implementation of environmental policies seriously.

The Lack of Economic Incentive Effect

As one of the important environmental management measures, economic incentives have been consistently emphasized by many countries' governments. During the last three decades, Chinese central and local governments have been promoting the establishment and practice of economic incentive policies, including the collection of pollution emission tax, tax reduction and exemption, accelerated facility depreciation, priority loans to environmental protection projects, foundations of environmental protection, emissions trading of air pollutant, and the collection of pollution compensation tax. However, most of these measures have yet to achieve the expected results. Mainly because of the inefficient policies and legal provisions, it turns out that the costs of environmental illegality are low, while the costs of environmental enforcement are high. It was estimated that the average cost of environmental illegality is less than 10 percent of the cost of treatment, and less than 20 percent of the cost of damage.

The Urban Environmental Issues Are Compounded

The original urban environmental issues have not been solved while new issues keep arising in China. First of all, the environmental pollution is increasingly outstanding around urban-rural boundaries. With the physical expansion of cities, the pollution issues of the environmental facilities which are located in suburban areas are quite severe, influencing the harmoniously development of urban-rural areas.

Secondly, the air pollution of many cities is compounded with coal-fired pollution and vehicle exhaust pollution. At present, China had already become the second largest vehicle consumption and production country in the world, the rapid development of vehicle industry had brought about huge pressure on the environment. The vehicle population growth has made the air pollution issues much more serious; some major cities' air pollutions are now transforming from coal-fired pollution to vehicle exhaust pollution. According to the statistics, the vehicle population in Shanghai is only 1/12 of it in Tokyo, but the total emissions of carbon monoxide, hydrocarbon and nitrogen oxide are relatively equal. In some major cities like Beijing and Guangzhou, more than 80 percent of the carbon monoxide and 40 percent of the nitrogen oxide are caused by vehicle exhaust. Haze weather in major cities has increased, and volatile organic pollutants have not been under control.

Thirdly, the problem of urban heat islands and urban desertification are serious, causing damage to natural ecological systems, and reducing the bearing capacity of the urban natural ecological system. Meanwhile the degradation of the urban natural ecological system has sharpened the contradictions between resource supplement and the social economic development.

THE STRATEGIES OF URBAN ENVIRONMENTAL PROTECTION

The first 20 years of the 21st century is the important strategic opportunity period of Chinese socioeconomic development and environmental protection. The comprehensive implementation of the scientific development concept, the rapid transformation of economic growth mode, and the construction of environmental friendly societies are important strategic tasks in China.

The Transformation of Economic Growth Mode

The transformation of economic growth mode, strict control of high-consumption and high-pollution industries is a fundamental approaches to achieve the resource saving and environmental friendly society.

Encouragement of the Use of Clean Energy

In order to achieve the goals of energy-saving and emission-reducing, increase the ratio of clean energy consumption, and encourage the utilization of renewable energy, the government should implement concessional loan and government subsidy policies to encourage the development and utilization of renewable energies for hydropower, solar and land-heat energy, and biomass energy. In accordance with air pollution issues, cities with better economic foundations should use clean energies such as natural gas and liquefied petroleum gas instead of coal. Beijing government has clamped down on the using of coal-fired boiler and promoted the using of natural gas boiler since 1998, and the air quality in Beijing has been improved dramatically.

Increase the Urban Environmental Infrastructure Construction and Operation level

The government plays the leading role in environmental infrastructure construction. In order to improve the urban environmental infrastructure construction and operation levels, it should emphasize the role of market mechanisms, for example, to promote diversification of environmental facilities investment, encouraging the joint-stock system, market-oriented operation and professional services.

Strengthening the Urban Environmental Risk Management

Along with the rapid development of the economy and society, the environment and ecological system is now under increasingly great pressure. In order to prevent severe environmental pollution events, the key points are improvement of environmental risk management, strengthening the study of urban environmental risk management, construction of comprehensive environmental risk management system, and the overall planning and rational distribution of high-risk industries such as the petroleum and chemical industries which are related to national economic livelihood. The improvement of environmental risk management and the construction of national environmental risk management policy system are strong guarantees of urban environmental protection.

Strict Law Enforcement

Central and local levels of the environmental protection departments are cabinet members of comprehensive management of the local environmental resources; they manage and protect urban environment by giving environmental protection strategies, implementing national environmental protection policies and regulations, and are involved in all comprehensive decisions of social and economic development. Governments need to strictly observe to the "Environmental impact assessment law of People's Republic of China,' to control the impact that economic development brings about on the environment when governments are making plans of land utilization, local construction, exploitation and utilization, and plans related to industry, agriculture, animal husbandry, forestry, energy, traffic, urban construction, tourism, natural resource exploitation, At the same time, governments should implement the environmental protection laws strictly, investigate and treat the law breakers severely, eliminate the wrong idea that the law does not punish multiple offenders, and protect the environment according to the law.

Formulating Economic Policy for Environmental Protection

Among all these implemented policies, some of them are inefficient because of the deficiency of incentives and motivations for enforcement, and some of their policy functions have been overestimated, and as a result they fail to meet their objectives. Thus, government should launch economic and motivated policies which are benefit to improve the control of pollutant emission.

Increase Science and Technology Investment

Along with rapid and continuous economic development in China, the environmental issues which appeared in developed countries' industrialization process during the last 100 years are now appearing mainly in China. Urban environmental issues are especially complicated, and the government must increase science and technology investment, to innovate the technology and management skills to solve the problems efficiently.

NOTES

1. Zhu Gengshen, *Environmental Management,* Beijing, China Environmental Science Press, 2002.
2. Wang Yongchen, *China Green Action*, New world Press, 2006.
3. Information Office of the State Council of the People's Republic of China, *Environ mental Protection in China,* Beijing, 1996.http://www.gov.cn/zwgk/.

4. Ministry of Environmental Protection of PRC, *Report on the State of the Environment in China 2006,* http://www.mep.gov.cn/zwgk/hjtj/.
5. Niu Fengrui Pan Jiahua Liu Zhiyan, *Urban Development in China: 30 Years (1978-2008),* Social sciences Academic Press (China), 2009.
6. Ministry of Environmental Protection of PRC, *Urban Environmental Protection in China,* http://www.mep.gov.cn/.
7. List of National Model Cities of Environmental Protection, http://www.zhb.gov.cn/cont/mhcity/.
8. In 2006, the standard reaching rate of drinking water was decrease, there were two reasons, first, the assessment measure was changed; second, added 88 county-level cities.
9. Information Office of the State Council of the People's Republic of China, *Environ mental Protection in China,* Beijing, 1996. http://www.gov.cn/zwgk/
10. Ministry of Environmental Protection of the People's Republic of China, *Report on the State of the Environment in China 2008.* http://www.mep.gov.cn/zwgk/hjtj/.
11. Li Yujun, Urban Environment Treatment in China, in *Urban Development in China: 30 Years (1978-2008)*, Niu Fengrui(et al, eds), pp.393-415, Social Sciences Academic Press (China),2009.
12. Ministry of Environmental Protection of the People's Republic of China, *Report on the State of the Environment in China 2007.* http://www.mep.gov.cn/zwgk/hjtj/.

Chapter 8

Ecological and Environmental Migration in China: Policy and Practice

Guoqing Shi and Jian Zhou[1]

As stated in the Stockholm Declaration on the Human Environment, "The human is not only the creature of environment, but also the shaper." The environmental problems are closely related to the survival and development of the human society. Specially, the conflict between human's needs for living and the production capacity of the environment becomes a hot topic which attracts worldwide concerns in the later 20th century. In China, ecological and environmental migration which appears since 1990s is the direct reflection of this conflict.

DEFINITION AND CONTENT OF ECOLOGICAL AND ENVIRONMENTAL MIGRATION

Definition

Ecological and environmental migration is defined as purposeful removal of polluting factories and their staff from cities or densely populated areas, transferring people out of environmentally and ecologically fragile areas, the socio-economic reconstruction of the relocated factories, communities and families, and the livelihood restoration activities for the people whose lives are affected by the limits and forced changes in their use of natural resources in order to protect and improve the ecological environment.[2] Ecological and environmental migration can be divided into two types, i.e., spontaneous and government-led migration. The latter is the human migration activities supported and organized by the government for the purpose of environment protection or improvement.

Content

The essence of ecological and environment migration is to solve environmental issues encountered by human beings by ways of population migration or transferring economic activities, and to coordinate the relationship between human beings and the environment.[3]

The targets of ecological and environmental migration are the human beings or economic activities associated with human. The reason for this kind of migration is damage or deterioration to the natural environment, which make it impossible for people to live there or engage in necessary economic activities. The impacts of ecological environmental migration span across social, economic, political, cultural, psychological aspects, and so on.

Differences between Ecological and Environmental Migration and Other Types of Migration

Migration can be divided into development induced resettlement, environmental migration, ecological migration, poverty-reduction migration, disaster forced migration conflict induced migration (refugees), economic migration and others. There are differences among these kinds of migration.

Different Causes

The cause of development induced resettlement is land acquisition and house demolition, which force people to move. The reason for ecological and environmental migration is damage or deterioration to the natural environment (air, water, soil, and voice) and so on, and the degradation of the ecosystem in desertification areas, areas with spreading of stony deserts, soil erosion areas, as well as nature (wetlands, animals, plants, rivers) conservation areas, flood storage and detention areas and so on. In these areas it is difficult to maintain species diversity of plants, animals, microorganisms or the integrity of ecosystem for human survival. Poverty-reduction migration aims to improve the living and production condition of the local poor people. The causes of disaster induced migration are natural disasters that lead to permanent or temporary loss of life and production conditions for humans living in their original areas. Conflict induced migration is the population movement in order to avoid losing life and property during wars or conflicts. Economic migration refers to the population movement which aims to improve people's economic and social conditions.

Different Purposes

Development induced resettlement is a kind of population movement caused by development projects, which include construction of water conservancy projects, electricity, roads, railways, airports, urban development projects, rivers, mine development projects, and economic development zones. These kinds of projects lead to land acquisition and house demolition activities, which cause population movement and socio-economic system reconstruction.

Poverty-reduction migration is a kind of trans-regional population migration which appears in China and aims to increase the share of resources, the level of income and standard of living of the rural poor population. According to "The White Paper of Rural Poverty Alleviation and Development in China (2001)," the state encourages and supports the poor farmers who live in the extremely poor living conditions to develop new ways to solve the food and clothing problems by migrating to other places.

Disaster induced migration is the kind of population migration as a result of natural disasters such as earthquakes, floods, tsunamis, landslides, drought, wind, rain, hail, locusts, volcanic eruptions, landslides, pestilence and others or the spontaneous population migration activities in order to avoid the impact of a disaster. For example, during the rehabilitation and reconstruction period after the "512 Wenchuan earthquake," around 100,000 urban and rural people need to be relocated to other places. After the flood of Yangtze River in 1998, a policy called "return field to Lake, return farmland to forests, build new towns for resettlers" was carried out, and 2.46 million rural people were relocated to other areas with adjustment of models of production and lives.

Conflict induced migration (refugees) refers to such a kind of population movement for avoiding the damage from crisis's or wars among nations, regions, tribes and groups, as a result of conflict over resources, economic, cultural, religious and other difficulties.

Economic migration is the population flow and movement among different cities, regions and countries, aiming to improve the production and living environment, socio-economic conditions and raise income, quality of life of individuals, families and groups.

Different Duration and Implementation Time

Development induced resettlement is time limited, and is normally finished within a planned time period in order to achieve the objectives and effectiveness of the project. Ecological and environmental migration generally can be completed within a long time, except when there is a major emergency; generally speaking, other measures can be taken to reduce the impacts on the environment. The impacts on ecosystems and the rehabilitation of ecosystems cannot be achieved in a short period of time; these measures usually require time flexibility for designing. Disaster induced migration takes place suddenly. A temporary relief needs to be conducted firstly. Migration during the period of post-disaster reconstruction also must be finished in a short time. Conflict induced migration also happens with a sudden and uncertainty of complete time. Poverty-reduction migration requires more consideration about the migrants' opinions on the proposed plans, and there can be a considerable period of time for design and implementation. Economic migration is entirely dependent on the will of the migrants themselves, as well as conditions, purpose, national and regional policy and institutional constraints, so its completion time is uncertain.

Different Needs and Sources of Funds

Funds for development induced resettlement come mainly from project budget. Funds for ecological and environmental migration come from public finance, society or the factories which bring pollution, and the individual families. Funds for poverty-reduction migration and disaster induced migration mainly come from the public finance, society and migrants' families. Funding for conflict induced migration mainly comes from the international community and migrants' families. Funding for economic migration mainly depends on migrant's own capital.

Different Voluntariness of the Migrants

For economic migration and poverty-reduction migration, migrants have the rights of choosing to move out or to stay. Such migration can defined as voluntary migration.

Ecological migration and environmental migration are both voluntary and involuntary. Voluntary migration refers to the population who has rights of choosing to move or to stay for improving and restoration of the ecosystem and environment. Involuntary migration refers to the forced restriction on economic activities of people, factories and communities, and movement of population, factories and communities for improving and restoration of the ecosystem and environment.

Development induced resettlement, disaster induced migration and conflict induced migration are involuntary because people should move to other places and rebuild the economic activities under the pressure of natural disaster, conflicts and wars.

Similarities between Ecological and Environmental Migration and Other Patterns

Except economic migration, ecological and environment migration shares several similarities with other patterns of migration: (i) generally there is population movement and transfer of economic activities. (ii) Living systems (supply of water, electricity, roads, medical services, schools and business), production systems (land and livelihood), economic system (employment and income) and social systems (social network and social organization) need to be reconstructed for individuals, families and even communities. The reconstruction requires necessary arrangement of policies, mechanism, institutions, finance, and planning and design before implementation. (iii) Complicated social, economic, environmental and ecological impacts will happen.

CAUSES OF ECOLOGICAL AND ENVIRONMENTAL MIGRATION

Migration Caused by Environmental Pollution

The movement of the population or economic activities caused by environmental pollution is called "environmental pollution migration." Sources of environmental pollution are mainly the following: the waste smoke, gas, water, residue and noise discharged from factories, the waste smoke, gas, noise, dirty water and garbage come from people's lives; the waste gas and noise from vehicles (all fuel vehicles, ships, airplanes etc.), water flow from farms where chemical matter is used in large-scale, such as fertilizer, pesticides, herbicides and other chemicals after irrigation, and waste water and residue from mines.[4] The air pollution, water pollution, soil pollution, noise pollution and light pollution caused by these sources have seriously affected people's daily production and life, and therefore lead to "environmental pollution migration."

Water Pollution

When a foreign substance comes into the water to the extent that original the purpose of water is damaged, water pollution occurs. Water pollution also refers to water quality deterioration caused by pollutants into water exceed the capacity of water self-purification. There are two kinds of ways which cause water pollution, including natural pollution and human element pollution. Mine exploitation brings water pollution and water and soil erosion nearby, so that the content of mineral in the water body is increased. The affected arable land and fish ponds may be damaged and the farmers have to transfer the production activities. Regular tiding makes residents in the affected areas to avoid the disaster by migration. Industrial "three wastes"[5] also leads to migration. For example, if untreated sewage from paper-making factories are discharged into rivers, the water quality of rivers will be seriously affected. As a result, the fishermen in the lower reaches of river will loose their livelihoods, and the source of irrigation water will be polluted so that the farmers' agricultural production will be affected, and effluvium caused by water eutrophication will affect residents in the surrounding areas. Therefore the waste water should be treated, or the factories need to be relocated. Residents living nearby can be relocated to other places with compensation. Meanwhile, a new waste water collection and treatment system should be set up. To protect the safety of drinking water and avoid water pollution in upper reaches of water sources area and surrounding areas caused by human activities, we need to restrict production activities of agriculture, industry, commerce, tourism, service industries and relocate residents and factories from the core and first-class protection areas. For example, the Qinhuai River in Nanjing was famous for "oar sound and light shadow" before the 1980's. However, it was seriously polluted in the years of 1980-1990. Nowadays it is attractive because of "blue waves people." In the treatment process of Qinhuai River, many shops and residents along the river had been moved into new places.

Air Pollution

If the content of some substances in the air exceed the allowed standard, it will be harmful for humans, animals and plants. There is also air pollution caused by equipment corrosion, construction corrosion, resources corrosion and decreasing air visibility and other problems.

The first way to solve air pollution problems caused by industrial factories is to relocate polluting factories. For example, in order to solve air pollution and water shortage in Beijing, China's Beijing Capital Iron and Steel Company was moved to Tangshan City of Hebei Province. The second way is that the residents nearby can be relocated in other places.

Noise Pollution

The main sources of noise are traffic noise (car horn, engine etc.), factory noise (textile factory, machinery industry, rock drilling machine, large fan etc.), construction noise and social noise and so on. The way to resolve noise pollution problem caused by industrial factories in densely populated areas is to move the polluting factories out of that area. For example, in China, many polluting factories in the central city areas are relocated to other places partly because of noise pollution.

Soil Pollution

This type of pollution refers to the soil pollutants - the "three wastes" material, chemical pesticides, micro-organisms etc. entering and accumulating in the soil, which cause changes of soil composition, structure and function, thus affecting the normal plant growth and development, resulting in decrease of crop yield and quality and negative impact on human health. For example, Japan's cadmium-rice event was due to soil pollution from mining which ultimately affected human health. Agricultural production activities may disappear because of soil pollution, and the people in the area may need to be relocated inter-regionally and their lives need to be restored.

Light Pollution

Light pollution, including white and bright pollution, artificial day pollution and day lighting pollution, is a new environmental problem. Light pollution declines visual acuity and increases cataract incidence; it makes it difficult to sleep at night, disrupts normal biological rhythms, leads to low energy, disrupts the balance within the human body, causes "light syndromes" such as dizziness, irritability, appetite loss, lack of energy and insomnia. If a building causes a serious light pollution, the only way to solve the problem is demolition or reconstruction to eliminate the impacts.

Migration Caused by Environmental Disaster

The population movement caused by environmental disasters such as geological environment (including mud-rock flows, landslides), and disasters caused by global climate and environmental change (floods, drought caused by rising sea levels) etc. is called as "environmental disasters migration."

Geological Disasters

These are typically disasterious environmental events due to natural and/or man-made geological changes, which make natural geological conditions deteriorate, and cause human life and property loses or seriously damage the resources and environment to support human activities.

Geological disasters are induced by many factors. The most prominent human-induced factors include the following. The exploitation of mineral resources damages the ecological environment and induces geological disasters to varying degrees. Damage to reservoir bank slopes after water storage with water conservancy and hydropower project may lead to a large number of landslides dilapidation, giving rise to serious geological disasters. The exploitation of original rock-foot slope in a variety of resource development projects destroys support of the mountain or creates new artificial slopes, causing the slopes to lose balance, resulting in landslides.

Various geological disasters have occurred and brought threat on people's life safety and property security that can not be ignored. According to incomplete statistics, in Hunan Province, since 1949, direct economic losses caused by three kind of geological disasters, large-scale landslides, collapse, mud-rock flows amounted to more than 10 billion Yuan, with more than 600 people died from the disasters.[6] In recent years, with increased human activities, geological disasters have also gradually increased. Earthquake, flood, tsunami and other natural disasters may cause geo-environmental disaster, leading to resettlement caused by geo-environmental disasters. For example, on May 12, 2008, degree 8 earthquake took place in Wenchuan, Sichuan Province in China, and brought immeasurable loss to the local residents. In order to reduce or avoid loss of life and property, the residents of earthquake disaster zone had to relocate. Subsequently, a number of waves of population transfer took place as part of a environmental disasters resettlement.

Disasters Caused by Global Climate and Environmental Change

Human migration and climate change have a close relationship. A large number of migrants appeared in some countries and regions as a result of these changes.

According to latest report from the United Nations Intergovernmental Panel on Climate Change (IPCC), humanity is faced with threat of sea-level rise caused by global warming. From New York to Tokyo, two-thirds of the world's cities with more than five million people may be flooded, affecting 600 million people.

In the report, scientists gave a series of alarming figures. Ten percent of the population of the world live at an altitude of less than ten meters. With climate warming, they face a huge threat to their survival. Among cities with more than five million people, nearly two-thirds are partly in the 0-10 meters above sea level. These cities include Tokyo, New York, Mumbai, Shanghai and Jakarta. Asia may be the most vulnerable areas in front of the threat of sea-level rise and flooding. Six hundred and thirty-four million people in the world live in low-lying. About three fourth of them live in Asia. Twenty-one countries, with more than half of the population are in this region. These include sixteen small island countries. A more serious problem is that some scientists warn that with the accelerated melting of the ice of Greenland and west Arctic Ocean, the speed of sea level rising will be faster and faster. An increase of more than one meter every one hundred years is highly likely. The IPCC Report predicts that according to the present population distribution, by 2080, about one billion people will face floods every year.

Sea-level rise has made a number of Pacific island countries suffer. Sea-level rise caused by global climate change may make the Pacific island nation of Tuvalu disappear from the map; the country's citizens have begun to be forced to migrate to New Zealand each year. Compared with many coastal cities in the West, China's coastal cities have some geographical advantage. They are usually not directly on the coast, but in the river estuaries. Shanghai and Tianjin are such cities. However, this does not mean that the Chinese coastal cities are very safe. If a storm surge and an astronomical tide appear at the same time, they may face great danger, such as in the 1998 Yangtze River floods.

Ecological Deterioration

Since 2000, ecological migration has become a focus of attention in China. In 1999, in order to protect and restore the ecological environment, the government of Zuo Banner of Alashan League implemented the "returning animal breeding grounds to forest and migration" project. In 2002, residents in Shennongjia National Nature Reserve were migrated outside so as to protect the wild animals. These cases were widely reported, and the concept of "ecological migration" was then gradually accepted by people. That period was also the start time of implementing the Western Development policies. Improving and protecting the ecological environment is one of the five basic policies. In the western areas[7] of China, ecological deterioration problems mainly include the following.[8]

The first problem is shortage of water resource and water loss and soil erosion. Western areas are drought areas and rainless. Annual precipitation on average is below 400 millimeters and in some areas is below 100 millimeters. In addition to shortage of water, water loss and soil erosion have also become serious problems. The total water loss and soil erosion area (exclude Tibet) in western area is 10,437 ha, accounting for 62.5 percent of the total loss and soil erosion area of China.

The second problem is land desertification. In 2004, the total land desertification area in western area was 2.5723 million square meters, which was 97.9 percent of the total land desertification area of the country. In Xinjiang Uigur Autonomous Region, the area of land desertification increases by 84.2 square kilometers each year.

The third problem is grassland degradation. In Inner Mongolia, Xinjiang, Gansu and Sichuan, overstocking has become a serious problem, with overgrazing by more than 40 percent commonly found on native pasture.

Ecological environment deterioration destroys a large amount of farmland and pasture. Meanwhile, excessive land reclamation and overstocking intensify water loss, soil erosion and grassland degradation, causing decline in the productivity of farmland and pasture. Moreover, an increasing population causes excessive land reclamation and leads to more serious ecological environment deterioration. Ecological deterioration and economic stagnation reinforces each other, creating a cumulative effect. Such a vicious circle, as showed in Figure 8.1, may undermine the foundation of economic development.[9]

According to the Chinese National Development and Reform Commission's research, before 2050, there will be 10 million people (primarily concentrated in the southwest and northwest and ecologically fragile areas) need to be moved out to solve the problem of ecological environment and rural poverty.

Figure 8.1 Mechanism of Ecological Poverty

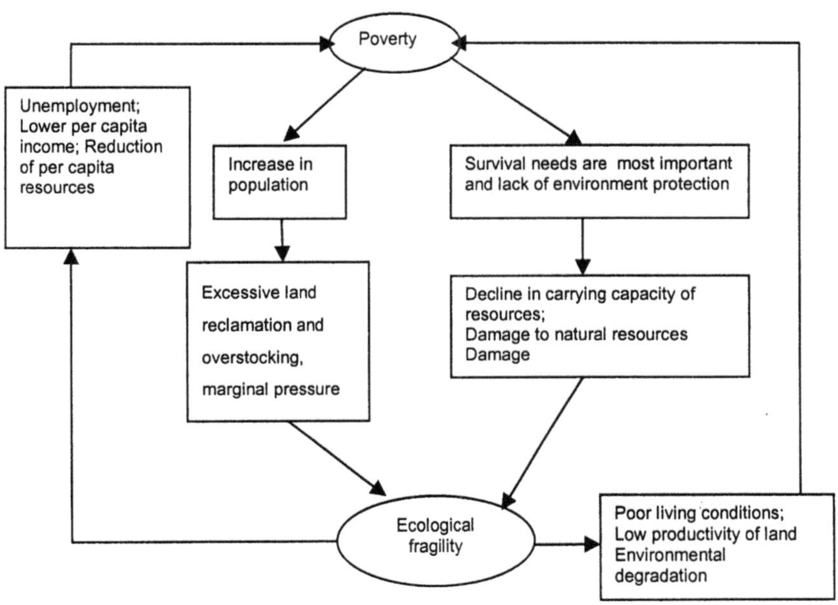

MODES OF ECOLOGICAL AND ENVIRONMENTAL MIGRATION

Ecological and environmental migration should be carried out according to different situations, such as the different levels of impact of the pollution or disasters, the feasibility of prevention, reduction and alleviation.

Relocation of Polluting Sources

Relocation of environmental pollution sources mainly refers to factories that produce waste water, exhaust gas or ambient noise. Its implementation ought to meet the following preconditions.

Tremendous Influence on the Area Caused by the Pollution Sources

Zero-discharge of the pollutants during production process is difficult to achieve due to limits in technology, finance or others reasons. If the pollution degree is at a low level, the pollution can be alleviated by nature's self-purification capacity. However, the environment will be greatly damaged if the level of pollution exceeds the environmental capacity of the region.

A Large Population Affected

As a result of urbanization and economic development, many factories are surrounded by an increasing population. Therefore more and more people's production and daily lives are being affected by pollution.

Possibility of Relocation of the Pollution Sources

As a precondition for relocation, a proper receiving area with suitable natural and social environments should be chosen for the relocation of the pollution sources. On the other hand, there should be enough financial and technical support for relocation and re-building.[10]

Case: the Relocation of Shougang Group

Shougang Group, a steel and iron group in China, will remove all its steel-making part from the Capital to Caofeidian in Tangshan city which is 225 kilometers away from Beijing. The new plant will be located on a man-made seashore. At present, the plants of SHOUGANG Group are only 17 kilometers away from Tian'anmen square so it is obvious that its pollution affects the air quality in Beijing to a great extent, especially in Haidian District. The inhalable particulate matter will be reduced by 18 thousand tons per year in Beijing because of the relocation, according to analysis from the official environment-protection departments. Even in Chaoyang District, the inhalable particulate matter can be reduced as a result. The relocation can also reduce 100 mg respirable particulate matter and 30 mg SO2 in Shijingshan District and some other areas. In order to reach Grade II of the ambient air quality, the total emission of all kinds of pollutants should be controlled between 33 thousand and 40 thousand tons per year. However, the emission at present is as high as 110 thousand tons. In spite of the .unavoidable discharge from civil use, transportation, infrastructure and some hi-tech industries, there is little leeway left for other pollution sources. Therefore, the total industrial emission should be greatly reduced. The steel-making part of Shougang Group has to be moved out of Beijing. Otherwise Beijing cannot make the air quality reach the standard nor carry out her promise of environment quality made when bidding for the 29th Olympic Games. As the case displays, the large range pollution by Shougang Group seriously endangers the Capital's environment, the citizen's health, and the Games with inestimable negative impacts. In the meantime, Caofeidian in Tangshan seems to offer an ideal location which impels Shougang's removal.[11]

Removal of the Affected People

The migration of affected population should meet the following preconditions.

High Cost of the Relocation of Pollution Resources

This refers to the situation when there would be high input by the factories on funds, human resources, technology and others, and the relocation would lead to high cost and waste.

A Smaller Population Affected

Less people means higher flexibility, less financial input and easier organization. This condition makes the removal more feasible.

Impossibility of the Relocation of the Pollution Sources

Various sorts of factors may make the relocation of the pollution sources impossible or insignificant. In this kind of circumstances, it is proper to leave the pollution sources and remove the population.

Case: the Villages Affected by Noises near BCIA

The construction of an extension to the Beijing Capital International Airport (BCIA) has cost 23 billion Yuan, with supporting projects costing several billion Yuan. The project is characterized with high input, long duration and advanced technology. Some villages were located only several hundred meters away from the runways. With the development of civil aviation and economy, the number of flights in the airport is increasing rapidly. Every week around 5000 flights depart the airport, according to statistics on the website of the BCIA. Villagers living nearby are disturbed by the huge noises. These villages affected by noises are now being gradually relocated. It is easy to understand that relocating the affected villages is much more feasible than to relocate the whole airport in this case.[12]

Governance on Treatment

In many circumstances, neither relocation of the pollution sources nor the removal of the population can be successfully carried out, since certain preconditions cannot be met. Then correct governance and treatment is taken into consideration.[13]

Re-Utilization of Resources

One of the most important methods of governance and treatment is resources re-utilization, which means the restoration of people's living with restriction of or forcible change of the method of resource utilization that has caused environment damage.

1. Fencing

This includes building fences in order to protect grasslands, forests and headwaters, such as fencing to prevent grasslands from overgrazing; building preservation barriers (examples: Helanshan Mountains preservation fence, Miyun Reservoir fence).

2. Removal

This means to remove the waters, people or animals that damage the environment, such as returning the aquaculture areas of Taihu Lake to lake and reorganizing and updating the surrounding pollution factories.

3. Transforming

This measure refers to changing the patterns of resource utilization, production and economic growth which go against environment preservation, such as controlling and transforming the planting and processing of yellow gingers in the upper reaches of Danjiangkou Reservoir.

4. Restricting

This measure refers to restricting people's and animals' activities which would worsen the environment, such as protecting reservoir headwaters by limiting resource exploitation, population increase, and tourism development.

Financial Compensation

1. Compensation for the environment

The producers, developers and operators who cause damage or harmful impacts to nature need to pay compensation for the environmental pollution, and pay the compensation to the future value which will be lost because of the current occupation on environmental resources. The methods for economic compensation include asking producers, developers and operators to pay credit funds, to pay taxes for contingent profits, environmental resources and pollutant discharge, to support environmental preservation projects carried out by the government, and to develop new substitution industries.[14]

2. Compensation for the affected population

The compensation for the population affected by environmental pollution is carried out through certain material or capital ways, including financial compensation and economic resources compensation, such as compensation to fishermen that suffer financial loss because of water pollution.

3. Case: Taihu Lake treatment

In 2003, the total GDP and financial revenue of Taihu Lake basin accounts for 13% and 19% of the nation. The region's urbanization rate has reached 60%; with only 0.3% land area of China, it supports 3% of the country's population. The problem of shortage of acceptable quality water is very conspicuous. The carrying capacity of the water is at a low level. Disputes caused by water pollution emerge endlessly; 70% of the water is in eutrophic condition, with blue green algae breaking out from time to time. These problems have seriously restricted the sustainable development of the basin. Governments of all levels now take a positive attitude towards the environmental problems and have formulated a series of policies of comprehensive administration on treatment, to ensure sustainable development. Since the region is in an advanced stage of economic development, rapid urbanization, with a dense population, and high dependence of fishery and local factories it is better to comprehensively govern the region by substantially reducing the aquaculture areas of the Lake, transferring the labor from aquaculture to urban areas and non-agriculture industries than to relocate the pollution sources or to remove the population.[15]

Ecological Migration

Ecological Migration is led by the government, mainly through immigration policies, by changing the population's production and lifestyles to reduce their dependence on the ecological environment in order to achieve ecological conservation.

At the national level, on December 14, 2002, the Sate Council promulgated the "Grain for Green Ordinance," which directly mentioned ecological migration. In Article 54, it is stated that during the implementation the "returning farmland to forests" project, the government encourages the implementation of ecological migration. Subsidies for living and production will be provided to the rural migrants.

Case: Ecological Migration along Tarim River Basin in Xinjiang Uigur Autonomous Region

In 2001, the Chinese Government invested seven billion yuan in environmental projects, including a comprehensive management system for the Tarim River ('cultivation discontinuation for enclosed forest restoration') and the resettlement of inhabitants, in order to remedy the environmental problems. Ecological migration was implemented in Luntai and Yuni Counties, Bayangol Prefecture, Xingjiang (a total area of 1,020,000 km^2).

This measure included reducing the use of of water by 1.5×10^8 m^3 from the Tarim River and transferring it downstream to enable water flow to the lower reaches of the river. This measure aims to reduce the destruction of natural poplar forests by people and livestock and to conserve water and improve water utilization efficiency to protect a total of 980 km^2 of natural poplar forests and shrub lands along a 100-km stretch of the river.

This area is intended to serve as a natural defense mechanism to protect the ecology and production of the natural poplar forests. A total of 2,420 people from 524 households from one township and two farms on the north shore of the Tarim River migrated to Layisu District in the Dina River Valley (part of Chunbake Township, Luntai County, 21 km southwest of the county center) and Lapa District (part of Dadaonan Township, Luntai County, 11 km southeast of the county center). After the migrations had been completed, all 40 km^2 of the land in Souko Township, Souko Agricultural Farm, and Kalchachi Livestock Farm were subjected to "cultivation discontinuation for forest restoration." Main policies for this ecological migration project are given below.

Living Facilities

Each household received a 2,667 m^2 housing plot. The government built the 80 m^2 brick-wood houses for each household, which included three bedrooms, one parlor and kitchen and bathroom. Ethnic minority's living customs have been taken into account to design the layout of the yard. The space between the outer wall and bounding wall of the yard is 5 m or 10m, so the residents can develop courtyard economy, such as planting vegetable and livestock breeding.

Production Arrangement

Agricultural-based resettlement is applied. A total of 3333 m^2 production land was allocated to each person with a thirty-year contract, in order to help them switch to farming from semi-herding and semi-farming. Meanwhile, the policies encourage the comprehensive development of the primary, secondary and tertiary industries.

Assistance Measures

Firstly, the government helped the people to move to the relocation site. The distance between the host areas and the original place is close to 100 km. Second, the government exempted some taxes and fees in the first year after the relocation. In addition, some subsidies were provided for agricultural materials and agricultural technology training.

Infrastructure of Relocation Site

Offices for the village committee, elementary school, clinic, cultural station, veterinary station, farm machinery station and credit cooperatives are all included in the plan. There is also special land for a new mosque.

Implementation of ecological migration will generate extensive effectiveness. On the one hand, according to a survey,[16] the per capita net income of the migrants was 2.5 times of the income of 2002. On the other hand, with the continued departure of inhabitants, and with large amounts of river water being transferred to downstream areas on five separate occasions, signs of life are appearing within the lower reaches of the Tarim River for the first time in over thirty years and natural vegetation restoration area reaches 180 square kilometers.

SEVERAL ISSUES TO BE STUDIED ON ENVIRONMENTAL MIGRATION

Environmental Migration is a new research field accompanied by global environmental problems and involves environment, population, resources, social, economic, political, psychological and many other disciplines. Different knowledge, theories, methods and techniques can be used from different perspective to conduct interdisciplinary research. The following preliminary issues are proposed.

The Sociological Perspective

From the sociological perspective, the following questions require further research: (1) migration activities' influences on society, culture, psychology, nation, gender, religion and resources; (2) identification, analysis, forecast, assess, control, management, and governance of social risks caused by migration activities and proposed control measures; (3) the livelihood, income, employment, land resource use of migrants; (4) the social and psychological problems induced by changes of living conditions and countermeasures; (5) migrants' social adaptation to community organizations, livelihood patterns, life styles, living habits, social environment, language, social interaction and so on; (6) social integration; (7) social network; (8) environmental migration patterns; (9) employment, income and social security issues for workers of relocated polluting factories; (10) formation mechanism, evolvement rules, analysis, forecast, evaluation, control, management and governance of social conflicts, social risks and crises triggered by migration; (11) social security such as employment ,health care, pension; (12) the relations between nation and society in migration activities; (13) the effectiveness of ecological and environmental migration activities for adjusting the relationship between people and nature.

The Management Perspective

From the management perspective, the following questions require further research: (1) the role, responsibilities, rights, duties and implementation of government; (2) the management of environmental migration, including five basic functions of management-planning, organization, command, coordination and control, and their execution before and during environmental migration implementation; (3) project management research, such as feasibility study on technology, economy, finance, environment and society etc, including project management research at different stages-survey, forecast, analysis, evaluation, decision-making, implementation, monitoring, inspection, evaluation after the project cycle; (4) the integrated management, risk management, time management, communication management and scope management of environmental migration; (5) the classification management at the stages of preparation and implementation for environmental migration; (6) the relationship and constraint between environmental migration and regional economic layout, social development strategy, and urban planning; (7) the factories' responsibilities and their fulfillment; (8) interest conflict and multi-game issues.

The Economics Perspective

From the economics perspective, the following questions need to be studied in the period of preparation ,transition, relocation, and reconstruction: (1) all types of tangible losses and benefits during migration activities, such as land (farmland, grassland and forest land), housing, production facilities, trees and so on; (2) losses and benefits on intangible assets, social capital, human capital in the process of migration; (3) external costs, such as the increasing production and living cost after relocation, the cost of local administration, and resources consumption in the process of transition and reconstruction; (4) losses and benefits of remaining residents in out-migration areas; (5) losses and benefits of communities and residents in relocated areas; (6) cost and benefit analysis of migrants' participation; (7) the loss and its evaluation to the stakeholder groups caused by environmental pollution; (8) the economic losses because of restrictions in resources development and utilization, agriculture, stock raising, industry, tourism, service industries and other productive activities in the process of environmental migration; (9) the standard of compensation and subsidy for the affected people; (10) internalization of the external cost; (11) direct or indirect impact on regional economic growth, employment, finance, revenue; (12) the financial subsidies and economic income from environmental migration projects; (13) theory and method for analyzing and evaluating the financial, economic, environmental, ecological and social cost and benefits; (14) land resource usage, value transfer and distribution of value-added.

NOTES

1. Paper supported by "Study on Social Security and Social Administration Caused by Land Requisition and Resettlement," a Key Research Project of National Foundation of Social Science (07ASH010), "Social Equity and Government Responsibility" (04JZD00017), a Key Research Project of Social Science Foundation by Ministry of Education of PRC, and "Study on Model in Ecological Migration Based on Development of Social Cohesion (05JGH187), Research Project of Ministry of Education of PRC.
2. Shi Guoqing, "*Involuntary Resettlement: Experience in the World and the Practice in China,*" presented at a workshop on Typologies of Relevance of the Study of Forced Migration, Washington, D.C., 2004.
3. Zhang Zhiliao, "Analysis of Covenant for Ecological Migration," *Journal of Chongqing University*, 2005 (8):149.
4. Mo Yingeng, *General on Environmental Science* (Beijing: Chemical Industry Press, 2009).
5. Waste gas, waste water and waste solid matters
6. Yang Shunquan and Li Zuohai, "Main Factors for the Geological Disasters in Hunan Province and Discussion on Countermeasures," *Engineering Geological and Hydrological*, 1999 (4).
7. Western area of China includes twelve provinces, which are Chongqing, Sichuan, Guizhou, Yunan, Guangxi, Shanxi, Gansu, Qinghai, Ningxia, Tibet, Xinjiang, Inner Mongolia. The total area is 5.38 million square meters, which is 56 percent of the total land area of the country. The total population is 0.287 billion, accounting for 22.99 percent of the population of China.
8. A. Aabulizi, *Study on Ecological Migration in Xinjiang*, China Economy Press, 2009..
9. Zhang Xiaoming, *Study on Ecological Migration in Western Area of China* (Xi'an: Northwest Agriculture & Forestry University Press, 2008).
10. Gou Xinyu and Xu Zhenghui, "Relocation of Shougang Group," *Economy Observation News*, 20 June, 2005.
11. Tang Zheng, "20 Years' Gaming for Relocation Shougang Group," *Beijing Daily*, 4 July, 2005.
12. Li Xiaojian, "6900 Farmers Moved Houses for Beijing Capital Airport," *Beijing Daily*, 16 August 2006.

13 . Xu Jihong and Gong Yan, "Discussion on Government's Management on Environment," *Journal of Shanxi Normal University*, June 2006.
14.Zhang Lei and Cheng Peng, "Government's Function for Ecological Environment Construction," *Forestry Economy*, 2002 (9).
15. Jiang Mingzhong, Xie Jianming and Zhuo Guorong, "Discussion for Promoting Development of Economically Weak Villages around Taihu Lake," *Jiaqing News*, 3 July, 2007.
16. A fifty household survey was conducted by the author in August 2008.

Chapter 9

"Cities Besieged by Garbage" and Countermeasures for Environmental Protection in Western Cities: A Case Study of Solid Waste Classification in Kunming

Zhang Jijiao, Li Yujun, Yin Peng and Guan Yi

BACKGROUND OF THIS RESEARCH

Solid waste[1] accumulation all over the world is increasing at the speed of 8 24 percent per year, 2.5-3 times of the speed of world economic growth. Urban areas, where most solid waste is produced, are at the same times the victims. Pollution caused by garbage, one of the public nuisances for cities in China is becoming more and more serious.[2] According to China's State Environmental Statistics 2000, the annual discharge of solid waste has reached 140 million tons and was increasing at the speed of 9 percent. The storage of solid waste has reached 6,000 million tons and covers an area of about 53,333 hectares. Among China's 668 cities, two-thirds have been besieged by solid waste.[3] For many years, garbage disposal has remained at a very low level, mostly because of open storage and burying, which seriously harms the soil, rivers, underground water and the atmosphere as well as human beings.

"Cities besieged by garbage" has become one of the most urgent and important problems for urban environmental protection.[4] The solid waste issue has been included in the *National 11th Five-Year Plan of Environmental Protection* as one of the central problems.

PURPOSE AND METHODOLOGY OF THIS RESEARCH

Recently, environmentalists and experts on garbage disposal have been dealing with urban solid waste issue from several perspectives—its harms, disposal techniques and rubbish dump construction. However, this issue has never been studied from the perspective of urban residents.

Through a case study of Kunming, capital city of Yunnan Province in western China, this research deals with urban solid waste issue. Appointed by the Ministry of Commerce as one of the first 24 pilot cities for renewable resources recycling system, Kunming government has started the project in 2006 and enacted *Regulations on Solid Waste Administration in Kunming*, and implemented *Regulations on Collecting Fees for Urban Solid Waste Disposal* on March 1, 2007. Despite the progress achieved, the problem of garbage disposal has becoming more and more serious due to the increasing solid waste and the lack of garbage disposal capacity. As an important part of building a resource conservation and environmental friendly society, solving the urban solid waste problem has been put on the agenda of urban construction in Kunming.

This research is not only of great practical importance to solve "cites besieged by garbage" problem in Kunming and to facilitate urban environmental construction in Kunming, but also of referential value for cities in western China to solve "cites besieged by garbage" problem.

This research project was initiated in September 2007, followed by 2 months of literature review, research outline and questionnaire design in October and November, field survey in December and 6 months of data processing, analysis, and report writing from January to June, 2008.[5]

During the field work, our staff collected 215 valid samples from environmental centers, garbage transit stations, main streets and communities, among which 80 are staff whose jobs are related to garbage disposal and 135 are non-staff whose jobs have nothing to do with garbage disposal.

AWARENESS, BEHAVIOR AND EVALUATION OF INTERVIEWEES ON SOLID WASTE ISSUE AND ITS CLASSIFICATION IN KUNMING

This study gathered first-hand materials through a questionnaire survey,[6] which included 6 aspects: (1) interviewees' basic information, (2) interviewees' awareness about solid waste issue and its classification, (3) channels for interviewees to get information on solid waste classification, (4) interviewees' behavior of solid waste classification, (5) interviewees' ways of dealing with separate collection and (6) interviewees' evaluation and suggestions on solid waste classification. Some parts of the statistical data in this research are listed below.

Interviewees' Awareness about Solid Waste Issue and Its Classification

Understandings of the Solid Waste Issue and Its Classification

1. Attitude on the Seriousness of Solid Waste Problem

In our survey, most (93.0%) of the interviewees realized that the solid waste problem is very serious in Kunming (see Table 9.1)

Table 9.1 Attitude on Solid Waste Problem in Kunming

	Valid	Percentage
Most serious	72	33.5
More serious	70	32.6
Serious	58	27.0
Not serious	11	5.1
Least serious	4	1.8
Total	215	100.0

Among those who chose "most serious," based on cross-table analysis by gender, age, education, income and occupation, opinions of different respondents are as follows:

In terms of gender, male interviewees exceed female interviewees by 3.2 percentage point; In terms of age, the percentage of those in the "41-50" age group are nearly 10 percentage point higher than that in other three age groups; In terms of education, the percentage (40-45 percent) of those whose educational levels are "below secondary school" and "above undergraduate" are 10-20 percentage point higher than interviewees from other educational levels; In terms of income, the percentage of those who had "no income" and those whose income was between "2,001-3,000" (RMB) yuan/per month are 8-18 percentage point higher than those in other income groups; In terms of occupation, percentage of "staff" whose jobs are related to garbage disposal is 16.4% higher than that of "non-staff" whose jobs have nothing to do with garbage disposal.

2. Awareness of the Promotion of Refuse Solid Waste Classification

Our survey in Kunming showed: over 70 percent of the interviewees expressed that they "know about" (76.2%) the efforts that Kunming government has undertaken in promoting solid waste classification.

Among those who "know about" solid waste classification work, based on cross-table analysis by gender, age, education, income and occupation, different interviewees hold different viewpoints:

The percentage of male interviewees is a little bit higher than that of females; Four age groups can be ranged according to percentage as "41-50" (90.7%), "26-40" (75.2%), "above 50" (72.4%) and "18-25" (62.9%); Connections can be seen between awareness and education. Percentage of "3-year college education" and above educational groups is 10 points higher than that of "high school" and below educational groups; Percentage of "2,001-3,000" (RMB) yuan/per month income group is the highest (90.2%), and that of "staff" whose jobs are related to garbage disposal are 12 percentage point higher than that of "non-staff" whose jobs have nothing to do with garbage disposal.

3. Evaluation on the Promotion of Solid Waste Classification

In Kunming, we found: over 80 percent of the respondents considered that the promotion of solid waste classification is far from enough (see Table 9.2)

Table 9.2 Evaluation of the Promotion of Solid Waste Classification

	Valid	Percentage
Most powerful	18	8.6
More powerful	18	8.6
Powerful	77	36.7
Less powerful	75	35.7
Least powerful	22	10.5
Total	210	100.0

Among those who chose “less powerful” and “least powerful”, based on cross table analysis by gender, age, education, income and occupation, opinions of different respondents are as follows:

Female interviewees (51.4%) exceed male interviewees (41.2%) by 10 percentage points; Different age groups can be ranged according to percentage as “26-40” (49.5%), “18-25” (48.6%), “41-50” (45.2%) and “above 50” (28.6%);

The higher the education level, the higher the proportion. Percentage of “graduate and above” group is 57.8% and that of “secondary school and below” is 33.3%; That of those who monthly income in “3,001-8,000” (RMB) yuan and “2,001-3,000” (RMB) yuan groups is higher (over 50%) than other groups; “Staff” exceeds “non-staff” by 8.4 percentage points.

4. Consciousness of Solid Waste Classification

The theory of planned behaviour (TPB) was used to guide an analysis of intentions to recycle household waste in a geographical area (Glasgow, Scotland) with relatively poor recycling facilities. A sample of 252 members of the public completed a questionnaire (response rate of 66%). In addition to TPB variables, the contributions of past recycling behaviour, perceived habit of recycling, and perceived lack of recycling facilities were considered. The TPB components contributed 29 percent to the variance of intentions to recycle; attitudes and perceived behavioural control (PBC) (but not the subjective norm) were significant on entry. Past recycling and perceived habit made significant independent contributions. [7]

Another research findings suggest that pro-recycling attitudes are the major contributor to recycling behaviour, and that these attitudes are influenced firstly, by having the appropriate opportunities, facilities and knowledge to recycle, and secondly by not being deterred by the issues of physically recycling (for example time, space and inconvenience). Previous recycling experience, and a concern for the community and the consequences of recycling are also significant predictors of recycling behaviour. The implications of the findings for the development and implementation of recycling schemes and for marketing and communication campaigns which advocate the use of these schemes are discussed.[8]

In our survey, over half (54.9%) of the respondents indicated that residents in Kunming are unconscious of solid waste classification, among which, based on cross-table analysis by gender, age, education and occupations, different interviewees hold different viewpoints:

The percentage of female (50% +) and male (60% +) respondents both exceed 50%; That of "18-25" and "above 50" age groups exceeds 40%, and that of "26-40" and "41-50" age groups is close to 50%; That of the "secondary school and below" educational level group exceeds 40%, "high school" over 50%, "3-year college education" 66.0% and "undergraduate and above" close to 70%; That of "below 1,000" (RMB) yuan and "1,001-2,000" (RMB) yuan income group is about 50%, "2,001-3,000" (RMB) yuan and "3,001-8,000" (RMB) yuan over 60%; That of "staff" and "non-staff" both exceeds 50%.

Awareness of Measures Taken by Government on Solid Waste Administration

1. Awareness of *Regulation on Refuse Administration in Kunming*

In our survey, over half (50.5%) respondents claimed that they "know" about *Regulation on Refuse Administration in Kunming* issued by Kunming Government, among which, based on cross-table analysis by gender, age, education, income and occupation, different interviewees had their own opinions as follows:

The percentage of female interviewees (48.1%) is less than male interviewees (53.3%); That of "41-50" age group is the highest (72.7%) while that of "18-25" is the lowest (29.4%); Different educational level groups can be ranged according to proportion as "secondary school and below" (55.1%), "3-year of college education" (51.9%), "graduate and above" (48.9%) and "high school" (47.7%); That of "2,001-3,000" (RMB) yuan income group is the highest (64.7%) while that of "3,001-8,000" (RMB) yuan is the lowest (31.3%); That of "staff" (77.5%) is higher than that of "non-staff" (34.3%).

2. Awareness of *Regulation on Charging Method of Garbage Disposal*

In our Kunming survey, over half (55.4%) of the respondents claimed that they "know" about *Regulation on Charging Method of Garbage Disposal in Kunming*, among which, based on cross-table analysis by gender, age, education, income and occupation, different interviewees hold their own viewpoints as follows:

The percentage of female interviewees (52.4%) is a little higher than male interviewees (58.1%); That of "41-50" age group is the highest (81.8%), and that of "18-25" is the lowest (20.6%); Different education level groups can be ranged according to percentage as "3-year of college education" (62.4%), "secondary school and below" (59.6%), "graduate and above" (51.1%) and "high school" (49.2%); That of "2001-3000" (RMB) yuan monthly income group is the highest (70.6%), and that of "3,001-8,000" (RMB) yuan monthly income group is the lowest (40.6%); That of "staff" (81.0%) is far higher than that of "non-staff" (40.3%).

3. Understandings of Solid Waste Management Policy and Its Classification

In our research, most (70.1%) respondents expressed that they had some knowledge about solid waste management policy and its classification (including those who "have a deep understanding", "have a better understanding" and "have some knowledge"), among which, based on cross-table analysis by gender, age, education, income and occupation, various interviewees have different opinions as follows:

The percentage of female interviewees (67.0%) is lower than male interviewees (75.5%); Different age groups can be ranged according to percentage as "41-50" (81.8%), "above 50" (76.0%), "26-40" (69.5%)" and "18-25" (57.2%); That of "3-year of college education" (76.9%) and "high school" (70.3%) is a little higher than that of "graduate and above" (68.8%) and "secondary school and below" (68.1%) in the different educational level groups; That of monthly income "2,001-3,000" (RMB) yuan group is the highest (85.7%); "Staff" exceeds "non-staff" by 32.8 percentage points.

Kunming Residents' Behaviors Regarding Solid Waste Classification

Classification of Solid Waste in Communities

In our 2007 survey, most interviewees (67.8%) realized that classification of solid waste has not been carried out in the communities they live in. It is worth noting that some respondents "don't know" (10.7%) whether separate collection of solid waste has been carried out in their communities, among which, based on cross-table analysis by gender, age, education, income and occupation, their different opinions are as follows:

The percentage of male interviewees (13.5%) is higher than female interviewees (7.5%); That of "18-25" (17.1%) and "above 50" (17.9%) is higher than that of other age groups; That of "3-year of college education" (13.2%) and "secondary school and below" (14.6%) is higher than that of other educational level groups; That of monthly income "below 1,000" (RMB) yuan group (15.5%) and "3,001-8,000" (RMB) yuan group (12.5%) are higher than others; That of "non-staff" (12.7%) is higher than "staff" (7.5%).

Classification of Solid Waste at Home

Source separation of household waste has gained popularity among the general public in Hong Kong. Despite an increasing number of married women involved in full time paid work, women are most involved in waste management in most

Hong Kong families. Thus, the attitude of this social group—housewives—is decisive in the success of the source separation of household waste and other recycling related activities. Attitudinal differences between housewives and the general public were found with respect to the amount of avoidable waste, the incentive level of waste collection charges, the acceptable level of a green premium, the support of source separation of household waste and a deposit-refund system on beverage containers and cell batteries.[9]

The household-recycling rate in the Borough of Burnley, England in 2001/2002 was only half the national average of 12%. A research employed both quantitative and qualitative surveys in order to ascertain whether householders' attitudes to recycling were contributory factors to the generally poor recycling performance and to investigate other social, cultural and structural influences. The Borough has a large Asian–British population concentrated in two deprived wards where recycling rates are particularly low, so special attention was given to ascertaining their attitudes towards recycling.[10]

1. Solid Waste at Home

In our 2007 survey in Kunming, respondents considered that there are mainly four kinds of solid wastes at home, "kitchen waste," "bottles and jars," "plastic bags" and "old newspapers and magazines."

Respondents of different ages have different opinions on main solid wastes at home. Those in "18-25" and "26-40" indicated "kitchen waste," "41-50" indicated "kitchen waste" and "bottles and jars," and "above 50" indicated "plastic bags." So did those of different incomes. Those in the monthly income "below 1,000" (RMB) yuan group and the "3,001-8,000" (RMB) yuan group indicated "kitchen waste," the "2,001-3,000" (RMB) yuan group indicated "plastic bags," and the "no income" group indicated "old newspapers and magazines". Majority of respondents from different gender and education groups indicated "kitchen waste" as the main solid waste at homes.

2. Classification of Kitchen Waste

During our survey in Kunming, most interviewees (70.4%) expressed that they "often" or "sometimes" collect kitchen waste separately. Among those who "often" collect kitchen waste separately, based on cross-table analysis by gender, age, education, different interviewees had their own answers as follows:

Percentage of male interviewees (41.6%) is higher than female interviewees (34.3%); The older the interviewees are, the higher the proportions are. Those in the "above 50" age group exceeds those in the "18-25" nearly 20 percentage points, "26-40" nearly 10 percentage points and "41-50" nearly 6 percentage points; That of the "secondary school and below" educational level group (52.1%) is the highest, "high school" (31.1%) and "3-year college education" (32.0%) level groups are the lower, and "graduate and above" (38.6%) in the middle. That of monthly income "below 1,000" (RMB) yuan group is the highest (45.5%), and that of monthly income "1,001-2,000" (RMB) yuan group is the lowest (28.6%); Respondents in "staff" exceed that in "non-staff" by over 20 percentage points.

3. Simple Classification of Solid Waste

During our survey in Kunming, most interviewees (84.8%) expressed that they classified solid waste simply at home, among which, based on cross-table analysis by gender, age, education, the interviewees had their different answers: Female interviewees (89.6%) exceed male interviewees (80.4%) by nearly 10 percentage points;

That of the "41-50" (40.9%) age group is the highest, "above 50" (17.9%) is the lowest, and "18-25" (31.4%) and "26-40" (32.4%) are in the middle; As education level increases, the percentage of those who expressed "never classify" decreases; The percentage of those in monthly income "3,001-8,000" (RMB) yuan group (59.5%) who expressed "keep on classifying" is the highest, while that of "no income" who expressed "never classify" is the highest; Respondents of "staff" who expressed "keep on classifying" exceed that of "non-staff" by over 10 percentage points.

4. Classification of Solid Waste in Public Areas

In our survey, nearly 70% of the interviewees (69.7%) thought that Kunming residents "can" or "sometimes can" classify garbage in accordance with the requirements in public areas.

Among those who expressed "can" classify garbage, male interviewees exceed female interviewees by 7 percentage points; the older interviewees are, the higher the proportion of those who expressed "can"; Interestingly, the greater the education level, the higher the percentage of those who thought that the public "can not" throw out garbage according to requirements in public areas;

The percentage of those in monthly income "2,001-3,000" (RMB) yuan group who thought the public "can" throw garbage is the highest (close to 50%). A higher percentage of "staff" (46.3%) considered that the public "could" throw garbage according to requirements in public areas.

Views on Promotion of Solid Waste Classification

Understandings on Urgency of Solid Waste Classification

In our Kunming survey, regarding the urgency of solid waste classification, most of respondents (83.7%) realized that it is urgent to promote solid waste classification. Although this is the government's business and duty, it should now be started with every individual (see Table 9.3).

Table 9.3 Urgency of Solid Waste Classification

	Valid	Percentage
Indeed urgent, and it should be started with every individual from now on	145	67.4
Indeed urgent, but it's the government's business and duty	35	16.3
Improving living conditions first, and solid waste classification second	20	9.3
Difficult to describe	9	4.2
Don't need to pay intentional attention	6	2.8
Total	215	100.0

Understandings on Difficulties in Solid Waste Classification

Our Kunming survey in 2007 showed: interviewees recognized four main difficulties in solid waste classification to be "weak individual consciousness" (26.4%), "has not yet become a common practice" (24.2%), "too few solid waste containers" (11.4%) and "lack of incentives" (10.7%). "Weak individual consciousness" and "has not yet become a common practice" are the top two.

Table 9.4 Reasons Why Solid Waste Classification is Difficult

	Valid	Percentage
weak individual consciousness	123	26.4
has not yet become a common practice	113	24.2
too few solid waste containers	53	11.4
lack of incentives	50	10.7
too much trouble	38	8.2
occupying too much time and energy	37	7.9
too little room at home	35	7.5
no return for individuals	16	3.4
others	1	0.2
Total	466	100.0

Regarding difficulties in solid waste classification: Interviewees of both different gender and occupation considered that the two most important difficulties are "weak individual consciousness" and "has not yet become a common practice." However, interviewees of different age groups had different opinions: those who are under 50 indicated "weak individual consciousness" and "has not yet become a common practice" as the top two, while those who are older than 50 indicated "no return for individuals" and "has not yet become a common practice."

Respondents of different educational level groups had different answers. As for the first choice, the "high school" and "3-year college education" groups indicated "weak individual consciousness," the "secondary school and below" group indicated "has not yet become a common practice," and the "graduate and above" group indicated "too little room at home." As for the second choice, the "high school" and "3-year college education" groups indicated "has not yet become a common practice" and the "secondary school and below" and "graduate and above" groups indicated "weak individual consciousness."

Interviewees of different income groups also have different answers. As for the first choice, those whose monthly income are between 1,001 and 3,000 (RMB) yuan indicated "weak individual consciousness," the "3,001-8,000" (RMB) yuan group indicated "has not yet become a common practice," the "below 1,000" (RMB) yuan group indicated "weak individual consciousness" or "has not yet become a common practice." As for the second choice, those in the "1,001-3,000" (RMB) yuan group indicated "has not yet become a common practice," the "3,001-8,000" (RMB) yuan group indicated "weak individual consciousness" and the "below 1,000" (RMB) yuan group indicated "too few solid waste containers."

Eagerness for Solid Waste Classification

Our Kunming Survey showed: over half (62.8%) of respondents expressed that they were "most eager" (39.0%) and "more eager" (23.8%) to carry out solid waste classification. Among those who expressed "most eager," the proportion of male interviewees (42.2%) is higher than that of female interviewees (36.2%); the proportion of those "under 50" age group (about 40%) is higher than that of the "above 50" age group (27.6%), and that of the "secondary school and below" educational level group (50.0%) is higher than that of the high school and above educational level groups (31-41%); that of the monthly "3,001-8,000" (RMB) yuan income group (29.0%) is lower than that of other income groups (about 40.0%), and that of "staff" (46.1%) whose jobs are related to garbage disposal is higher than that of "non-staff" (35.1%) whose jobs have nothing to do with garbage disposal.

Appraisal and Suggestions on Solid Waste Classification in Kunming

Performance of Solid Waste Classification in Kunming

During our survey in Kunming, most interviewees (58.2%) gave a modest evaluation of the performance of solid waste classification in Kunming, stating that it was neither too good nor too bad.

Table 9.5 Performance of Solid Waste Classification

	Valid	Percentage
better	14	6.6
good	18	8.5
ordinary	124	58.2
bad	44	20.7
worse	13	6.1
Total	213	100.0

With regard to "better" and "good" performance of solid waste classification in Kunming, different interviewees' answers are as follows:
The percentage of male interviewees (19.2%) is higher than female interviewees (10.4%); that of the "41-50" age group (20.9%) is the highest, while that of the "18-25" age group (2.9%) is the lowest; that of the "high school" educational level group (22.0%) is the highest, while that of the "3-year college education" educational level group (7.5%) is the lowest; that of the monthly income "2,001-3,000" (RMB) yuan group (20.0%), and the "no income" group (22.2%) is higher, while that of the "3,001-8,000" (RMB) yuan group (6.3%) is the lowest; "Staff" (24.3%) exceeds "non-staff" (9.6%) by 14.7 percentage points.

Main Difficulties in Solid Waste Classification

In Kunming, among the nine difficulties listed, the top three are "public environmental behavior" (17.2%), "lack of legislation support" (14.9%) and "lack of incentives for local residents" (14.7%).

Table 9.6 Main Difficulties in the Promotion of Solid Waste Classification (multiple-choice)

	Valid	Percentage
public environmental behavior	89	17.2
lack of legislation support	77	14.9
lack of incentives for local residents	76	14.7
lack of cooperation from the public	67	13.0
lack of facilities	63	12.2
lack of effective measures	47	9.1
less investment	46	8.9
shortage of working staff	27	5.2
lack of incentives for staff	23	4.5
Others	1	0.2
Total	516	100.0

With respect to the most difficult aspect in the promotion of solid waste classification, the interviewees of different age, education, income and occupation have different views as follows:

Table 9.7 Interviewees of Different Ages and Difficulties (multiple-choice)

	18-25		26-40		41-50		Above 50	
	Valid	%	Valid	%	Valid	%	Valid	%
lack of legislation support	11	13.4	38	14.8	20	18.2	8	12.7
lack of cooperation from the public	15	18.3	33	12.9	13	11.8	6	9.5
lack of incentives for local residents	12	14.6	43	16.8	10	9.1	11	17.5
lack of incentives for staff	4	4.9	14	5.5	2	1.8	3	4.8
lack of facilities	12	14.6	30	11.7	10	9.1	9	14.3
less investment	4	4.9	21	8.2	14	12.7	6	9.5
public environmental behavior	13	15.9	46	18.0	19	17.3	9	14.3
shortage of working staff	2	2.4	12	4.7	6	5.5	7	11.1
lack of effective measures	8	9.8	19	7.4	16	14.5	4	6.3
others	1	1.2	0	0.0	0	0.0	0	0.0
Total	82	100.0	256	100.0	110	100.0	63	100.0

Among interviewees in different age groups, those in "18-25" age group indicated "lack of cooperation from the public" (18.3%), The "26-40" age group indicated "public environmental behavior" (18.0%), the "41-50" age group indicated "lack of legislation support" (18.2%) , and the "above 50" age group indicated "lack of incentives for local residents" (17.5%).

Among those in different educational level groups, the "secondary school and below" group (15.7%) thought "lack of cooperation from the public," the "high school" group (18.5%) thought "local residents," both the "3-year of college education" group (17.4%) and the "graduate and above" group (21.6%) thought "public environmental behavior."

Among those in different monthly income groups, the "1,001-2,000" (RMB) yuan, "2,001-3,000" (RMB) yuan and "3,001-8,000" (RMB) yuan groups thought "public environmental behavior," while those in the "below 1,000" and "no income" thought "lack of cooperation from the public." Among those in different occupation groups, "staff" chose "lack of legislation support" (16.2%), while "non-staff" chose "public environmental behavior" (19.1%).

Opinions and Advice on the Improvement of Solid Waste Classification in Kunming

1. Opinions on Effective Measure to Reduce Solid Waste

In Kuming, all interviewees considered the top three measures to reduce solid waste to be "encourage to use recyclable package materials" (26.1%), "avoidance of over-packaging" (16.2%) and "less plastic bags, more cloth bags" (16.2%). Of which, the most effective one is "encourage use of recyclable package materials."

Table 9.8 Effective Measures to Reduce Solid Waste

	Valid	Percentage
encourage to use recyclable package materials	135	26.1
avoidance of over-packing	84	16.2
less plastic bags, more cloth bags	84	16.2
separate collection of solid waste	67	12.9
increase publicity and education	61	11.8
more environmental friendly products	43	8.3
charge for plastic bags in stores	23	4.4
high fees for solid waste	20	3.9
others	1	0.2
Total	518	100.0

2. Advice on the Improvement of Solid Waste Classification

During our survey in Kunming, interviewees considered that government should take measures on the following four aspects: "increase publicity and education" (29.3%), "separate collection of solid waste at communities" (19.7%), "more garbage containers in public areas" (17.3%) and "adopt incentive policies" (12.9%).

MAIN PROBLEMS OF GARBAGE DISPOSAL IN KUNMING

First, the lack of garbage disposal capacity. Kunming consists of a north district (downtown area), an east district, a south district and a west district, up to 2020, it will cover an area of 460 square kilometers with a population of 4.5 million, several times that of the present size, an area of 180 square kilometers with a

Table 9.9 Improvement of Solid Waste Classification (multiple-choice)

	Valid	Percentage
increase publicity and education	161	29.3
separate collection of solid waste at communities	108	19.7
more garbage containers in public areas	95	17.3
adopt incentive policies	71	12.9
charge by amount of solid waste	35	6.4
more solid waste incineration plants	28	5.1
more vehicles for garbage classification	28	5.1
more solid waste landfill sites	23	4.2
Total	549	100.0

population of 2.4 million. With the volume increase, it is estimated that daily discharge of garbage would reach 7,000 tons, which would lead to a widened gap of garbage disposal. Demands for garbage disposal facilities will rise because environment and ground water are being polluted due to large garbage dumps caused by the lack of regulated disposal facilities.

Second, unreasonable distribution of facilities. So far, over 90 percent of garbage disposal facilities are used to deal with urban solid waste, while rural garbage is mainly processed by way of simple landfill, which doesn't meet the standard of harmlessness. With the inclusion of rural garbage into urbanized administration, garbage disposal capacity should be improved and distribution of solid waste disposal facilities should be readjusted as soon as possible.

Third, single technique of garbage disposal, and low level of resource recovery and reduction. At present, garbage disposal in Kunming is mainly by way of landfill, which takes too much land with a very low level of recovery and reduction. Direct landfill of crude waste with rich organic compounds and moisture contents might cause many problems, including potential risks in garbage dumps and other waste, all of which would pollute the surrounding environment seriously.

Fourth, decreasing land for facilities construction and difficult choices for facilities locations. With urban development, Kunming is suffering from a shortage of land for urban construction. Apart from that, the improvement of utility institutions of land and the raising of environmental requirements also increase the difficulty of choosing facilities locations. Simply land filling would not only waste land resources, but also increase the cost of garbage disposal.

Fifth, imperfect system of policy support. On the one hand, shortage of garbage disposal fees, caused by an imperfect charging system, various charging standards, and a low rate of garbage disposal charges, is not conducive to the formation of a diversified investment of public funds into garbage disposal. On the other hand, it is difficult for such construction to choose locations, meaning that urban garbage disposal facilities all locate in remote rural areas, which harms the local environment. Therefore, relevant environmental compensation systems should be created as soon as possible.

CONCLUSION: ANALYSIS AND SUGGESTIONS

Several suggestions are put forward regarding the current situation of solid waste administration and Kunming residents' opinions and advices.

First, regulations and legislation on solid waste classification should be improved and economic incentives should be established. Regulations and legislation, which should be constantly improved to regulate the public's behavior of solid waste classification, are more important than the environmental consciousness of the public. At the same time, complementary implementation mechanisms should be strengthened. Without these mechanisms, regulations and legislation are but empty shells. "Failure to abide by the law" is worse than "no law to abide by." The enforceability of implementation mechanisms is determined by penalty cost. Powerful enforcement, which raises the penalty cost, will make any breach of regulations unworthy, namely, the penalty cost is higher than benefit. Corresponding punishment should be inflicted on those who break the regulations such that their violations are made unproductive. Instead, material incentives should be awarded to those who do well in solid waste classification. Effective regulations and economic measures can be of use to regulate and restrict the publics' behavior.

Second, relevant systems should be created to regulate garbage discharge at fixed times and fixed places. It is inefficient for the public to form the habit of separate collection of solid waste using separate containers are placed at public areas for residents to classify by themselves. Instead, it would be much more favorable for sold waste to be discarded at fixed times and places so that people could follow the unified standard to put different types of garbage into relevant rubbish bags, and then, these classified solid waste could be collected, transferred and disposed separately. On the one hand, it is conducive to reduce garbage production that residents purchase unified classification bags to pay a certain amount of waste disposal costs. On the other hand, residents' behavior regarding garbage classification can be standardized and mixed delivery can be reduced through the supervision of specialized staff. This mechanism is conducive not only to the reduction of solid waste production and costs, but also to the mixed delivery under public discard system.

Third, mechanisms should be set up to promote solid waste classification information. According to our field survey, the respondents did not quite understand the garbage classification method recommend by the government, which shows that the government's promotion should be strengthened. A constant promotion system should be set up to disseminate knowledge and policies of garbage classification and to provide residents with correct guidance. Publicity of garbage classification system of knowledge would enhance residents' awareness of environmental protection and enthusiasm to participate in garbage classification, and to equip them with correct knowledge regarding garbage classification.

Fourth, pilot tests for solid waste classification should be promoted. Through our survey in Kunming, we found a remarkable effect of solid waste classification that the final amount of waste in need of disposal was dramatically reduced. Therefore, it is of great significance that pilot tests are carried out for the administration of separate collection of solid waste.

Our survey in Kunming shows that there is more than half (62.8%) of the respondents were "the most eager" (39.0%) and 23.8% of them were "more eager" to express their welcome to implement the solid waste classification. Generally speaking, the vast majority of people are very keen to participate in the separate collection, starting from their own and protecting the environment. Therefore, further expansion of community-based pilot garbage classification is very necessary.

The garbage management system should be followed to avoid mixed transfer of solid waste after separate collection. According to our survey, most urban residents are aware of solid waste classification, but their participation rate is relatively low. Therefore, we must pay special attention to attract more Kunming residents to participate in solid waste classification.

NOTES

1. The terms "solid waste," "garbage" and "rubbish" are used interchangeably in this chapter.
2. Li Yujun, "Solid Waste, Urban 'Cancer,'" in *Science and Technology in China (Zhongguo Keji Zongheng)*, Vol. 9, 2004.
3. Ministry of Environmental Protection of the People's Republic of China, *China's State Environmental Statistics 2000*.
4. Li Yunjun, "Discussions on the Improvement of Urban Solid Waste Administration in China," in *Journal of the Fujian Provincial CPC Committee Party School*, Vol. 4, 2008.
5. This chapter is a part of a research project report. This project was formally initiated in September 2007 as "A Key Project of Study on National Affairs in CASS," which finished with a research report of nearly 300 pages. Professor Zhang Jijiao, who was in charge of this project, wrote this chapter in Chinese, which was translated into English by Yin Peng. Li Yujun, with her research experience in garage classification of urban areas in Beijing and Shanghai in China and five cities of Japan, participated in the design of framework and questionnaire. Data processing and description were finished by Yin Peng and Guan Yi. We would like to express our thanks to our colleague Prof. Du Fachun and four master students of Prof. Gao Dengrong in Yunnan University for Nationalities, Zhang Liyin, Li Jun, Zhu Jiangning and Shi Ling, who participated in field research in Kunming.
6. The questionnaire of this project was designed on the basis of that of "Beijing Solid Waste Planning" presided over by Li Yunjun with some appropriate modifications.

7. Christina Knussen, Fred Yule, Julie MacKenzie, Mark Wells, "An analysis of intentions to recycle household waste: The roles of past behaviour, perceived habit, and perceived lack of facilities," in *Journal of Environmental Psychology*, 2004, Vol. 24, pp.237-246.

8. Michele Tonglet, Paul S. Phillips, Adam D. Read, "Using the Theory of Planned Behaviour to investigate the determinants of recycling behaviour: a case study from Brixworth, UK," in *Resources, Conservation and Recycling*, 2004, Vol. 41, pp.191-214.

9. Shan Shan Chung, Chi Sun Poon, "The Attitudinal Differences in Source Separation and Waste Reduction between the General Public and the Housewives in Hong Kong " in *Journal of Environmental Management*, 1996, Vol. 48, pp.215-227.

10. M. Martin, I. D. Williams, M. Clark, "Social, cultural and structural influences on household waste recycling: A case study," in *Resources, Conservation and Recycling*, 2006, Vol. 48, pp.357-395.

Chapter 10

Ecological Migration Practices and the Placement of Migrants and Its Impact on Poverty Alleviation and Development: A Case Study of the Huanjiang Maonan Autonomous County Guangxi

Fang Sumei

INTRODUCTION

Ecological migration is an important part of many countries' environmental policies. It is generally believed that the ecological migration policy of contemporary China began in the early 1980s, and was firstly implemented in the southern mountainous areas of the Ningxia Hui Autonomous Region. Due to extreme environmental deterioration, the residents the region were not guaranteed a normal life. Beginning in 1982, residents began migrating to other places under the organization of the Chinese government.[1] These policies have since spread to many other regions in China. As a result, it has become an important part of the Chinese government's ecological immigration policy. Since these policies were among the Chinese government's early anti-poverty program and are an important way to finally eliminate absolute poverty in the rural areas, they are of important realistic significance.

In the process of implementing an ecological migration policy, a model that combines immigration placement with poverty alleviation and development and was adopted by the Guangxi Zhuang Autonomous Region has achieved relatively significant results. Guangxi Zhuang Autonomous Region is a region with a larg population living in poverty. In the early 1990s, five million people lived below the poverty line in Guangxi, of which 80 percent lived in the rocky mountainous areas (i.e., Karst areas) with inadequate water, poor infrastructure, and few resources. In 1993, the Guangxi Zhuang Autonomous Region Government decided to migrate and place 250,000 poverty-stricken people from the rocky mountainous areas over seven years. By 1998, more than 200 immigration placement settlements had established in Guangxi Autonomous

Region, and twenty million people from 22 poor, mountain areas had been migrated to new areas to engage in comprehensive agricultural development.[2]

Among the immigration development projects implemented in Guangxi, Huanjiang Maonan Autonomous County has played an important role. Huanjiang Maonan Autonomous County, established in 1986, is the only Maonan autonomous county in China. Most western and northern regions in that county belong to karst landforms; large numbers of people from minority groups, such as the Maonan, Zhuang and Miao, live in poverty. As early as the early 1950s, the Huanjiang County People's Government mobilized people in the mountainous areas to migrate to areas with relatively flat terrain and relatively good farming conditions, in combination with the land reform movement. In the 1980s, the Huanjiang County Government began to organize more groups from mountainous areas to migrate to the plains and hilly areas, and carry out ecological migration practices that were combined with migration placement and poverty alleviation and development. At the same time, from 1993 to 2003, Huanjiang County accepted nearly 60,000 migrants from the Yao, Han, and other minority nationalities who lived inmountainous areas such as Donglan, Dahua, and Duan.[3] All these migration placement, poverty alleviation, and development activities in Huanjiang County have certain common features that may be of interest to ecological migration practices of Guangxi and even the whole country.

During my fieldwork in Huanjiang County in April 2008 and from January to February 2009, I investigated local ecological migration work, including visits to a poverty-stricken village in the rocky mountainous area named Xiyuan in Xianan Township, two migrant settlements named Dongyutun and Neitong in Changmei Township, and Chenshuang Migrant Site in Chengbei Community, Shaiguling Migrant Site in Chengnan Community, Sien Town near the county seat. It should be noted that my ancestral home was in Gedan Village, former Shangnan Township at the rocky mountainous area in Huanjiang. In the early 1950s, my second uncle and his family immigrated to Pingyuan Village, Luoyang Town. My mother reclaimed wastelands and cultivated lands with my second uncle and his family for one year in 1955. So far, the third daughter of my second uncle and her family still live in Pingyuan Village, which has become a major migrant farm in Huanjiang Maonan Autonomous County. During my interview process, I received assistance from my parents, the fourth daughter of my second uncle, Mr. Pan from Sien Town People's Government, and Mr. Mo from Chengbei Neighborhood Community, Sien Town. They provided me not only with their personal knowledge but also valuable background and contacts.

In addition, Mr. Qin Yonggui who now works for Luoyang Town People's Government, Huanjiang County once was the manager of the Chenshuang Migrant Site. In December 1996, he led 78 villagers (one person from each household) from Shangnan Village, Gedan Village, Shandong Village and Minquan Village, Shangnan Township to the Chengshuang Migrant Site to start reclamations and construction. He was stationed there for one year, and witnessed the establishment and construction of Chengshuang Migrant Site. His recollection and description provided first-hand material for this study.[4] It should be noted that all the data and materials used in the study are from documents, reports, summaries, reporting, statistics and profiles and the like obtained during the field investigation as well as a questionnaire and my notes.

The model combining immigration placement and poverty alleviation development that was implemented in Huanjiang Maonan Autonomous County is a representative example of Chinese ecological immigration policy and its effects are obvious. However, after twenty years of practice and testing, some problems have been exposed or produced. I think that scientific summary research on the immigration placement and poverty alleviation practices in Huanjiang Maonan Autonomous County may provide an objective understanding and evaluation of Chinese ecological migration policy from one perspective.

REASONS AND PROCESSES OF MIGRATION DEVELOPMENT IN HUANJIANG

There is a large area of typical karst landform in the area adjoining Guangxi, Guizhou, Yunnan region and provinces in southern China, that has high mountains and very beautiful scenery. Among them, Hechi City and Baise City located in the northwestern Guangxi Zhuang Autonomous Region are the main karst mountainous areas of Guangxi. In this region, peak clusters, peak forests, depressions, and underground rivers are extensively developed, and the exposed limestone area accounts for 39.6 percent of the total region (mountainous areas account for 48.096 percent of Hechi City, and 31.2 percent of Baise City).[5] Due to special hydrodynamic conditions, geological structure conditions, and climatic conditions, as well as the slow soil formation rate and thin and uneven soils and special chemistry and physics, soil erosion is increasing while soil fertility and the environment are declining. In the early 1990s, there were about 500,000 people living in the rocky mountainous areas, with an average per capital cultivated land rate of less than 0.02 hectares (equivalent to 18.87 percent of the national per capita cultivated land and 8 percent of the world per capita cultivated land), and the living conditions were very difficult.[6] This region is still one of the poorest regions in China.

Huanjiang Maonan Autonomous County is an important part of karst areas in northwestern Guangxi. It is located in northwestern Guangxi and is southeast of the Yunnan-Guizhou Plateau. Its geographical location is 107° 51' E-108° 43' E and 24° 44' N-25° 33' N and is under the jurisdiction of Hechi City. In terms of topography, northern part of the county has a higher elevation than the south. It is surrounded by mountains, and the south-central part is hilly and generally appears to be a basin. Most of the western and northwestern parts are high mountains and most of northeastern part is mountains with few trees; the altitude in the north is 1000-1600 meters above sea level and that in the south is 200-800 meters. The land area of the county is 4,557 square kilometers, which ranks sixth in Guangxi.[7]

Huanjiang Maonan Autonomous County was established on November 18, 1986 with the approval of the State Council. It is the birthplace of the Maonan people, one of twenty-two Chinese ethnic groups with small populations and where most Maonan live. The area is generally known as "Maonan Mountain Villages." In 2007, the county had jurisdiction over 6 towns, 5 townships and 6 ethnic townships including 148 administrative villages (communities), 2,791 natural villages, and immigration farms and settlements with a total population of 405,800 people, of which the agricultural population accounted for 90.6 percent of the total population. Thirteen ethnic groups, including the Maonan, Zhuang, Yao, Miao, Dong, Mulao, and Shui, live in Huanjiang Maonan Autonomous County. Their total population is 378,000 people, which accounts for 93.17 percent of the county's total population. Of this, the Maonan population is 59,000, which is 14.5 percent of the total population in the county. They live in 10 townships and towns including 72 administrative villages and 1300 national villages, and covers an administration area of 2,918 square kilometers, accounting for 63.8 percent of the total area of the county.[8]

In this area of rich resources and beautiful sceneries, but with continuous mountains and poor infrastructure, the agricultural conditions for poor. In particular, the agricultural conditions in townships and towns such as former Shangnan, Xianan, Mulun, Minglun, Dongxing, Longyan, Changmei and Xunle located in the western and northern parts of the county are extremely poor because they are in rocky and semi-rocky mountains with barren land, limited sunshine, and a shortage of water source. Thus, for many years, the economic and social development of Huanjiang Maonan Autonomous County lagged behind other areas and a farmer's life was extremely difficult. After reforms and looser economic controls, the economy and society of Huanjiang Maonan Autonomous County has rapidly developed and people's living standards have gradually improved. However, due to the restrictions of the natural conditions and the influence of historical factors, Huanjiang Maonan Autonomous County is still unable to completely shake off poverty and a lack of development.

In the 1980s, before the establishment of the Huanjiang Maonan Autonomous County, it was designated one of twenty-three impoverished counties in Guangxi. In February 2002, Huanjiang Maonan Autonomous County was listed as a key county for poverty alleviation and development for the new planning period by the State Council. In the end 2006, 100,900 low income people and 18,400 people below the poverty line (per capita annual income is less than RMB 683 Yuan) in the county. Furthermore, 880 villages in the county were still difficult to access and more than 200 villages did not have electricity, and more than 138,100 people had limited access to drinking water (the annual water shortage period is 4 months and above).[9]

However, according to a land resource survey in the early 21st century, there are 80,414.55 hectares of unused land in Huanjiang Maonan Autonomous County, which ranks No. 1 in northwest Guangxi; the area is an important land reserve northwest Guangxi. Of the unused land, 76.13 percent is high quality grassland. This region has some of the richest and most concentrated cultivated land in Guangxi and plays an important role in the occupation-supplement balance of the regional cultivated land. Most of the unused grassland resources have good conditions, are less than 50 meters above sea level with a slope of 6°-15°, and have good water conditions. There are no significant obstacles to agricultural production and the soil's fertility can be improved quickly. Further, the Dahuaijiang and Xiaohuanjiang Rivers and their tributaries run through the region, and include 239 large and small reservoir projects with a capacity of 11,990,000 m^3, so the irrigation conditions are good.[10] Favorable light, heat, and water are very beneficial to crops like rice and sugar cane, and can meet the requirements of 2-3 harvests a year. It can also support crops that require warm temperatures such as corn, peanut, soybean, sweet potato, jute, and ambary hemp with favorable conditions. All of these products support the basic conditions for ecological immigration.

As early as in the early 1950s, while promoting land reform, the Huanjiang County People's Government mobilized farmers living in mountainous areas to migrate to grasslands suitable for farming reclamation. My second uncle and his family moved from Gedan Village, Shangnan Township to Pingyuan Village, Luoyang Town to reclaim land in 1955. My mother, who was newly married, also went to work with them for a year. However, because two of my second uncle's children died from illness, they worried about Fengshui, so a few years later, they moved back to their former home. It is said that most of the villagers who went with them stayed in Pingyuan. This was the beginning of farmer migration in Huanjiang County, due to poverty after the founding of the People's Republic of China. Since the scale was not big, the government only provided local mobilization support and land allocation.

In the late 1980s, the problem of rural poverty garnered the government's serious interest. For poverty-stricken people in rocky mountainous areas, there was no other better positive solution than ecological migration. Therefore, the Huanjiang Maonan Autonomous County People's Government actively promoted poverty alleviation through economic development. It decided to let poor farmers move to the slope areas for comprehensive agricultural development. In 1989, it used the large areas of unused slopes in the original state-owned Pingyuan Livestock Farm to establish the first economic entity for migration placement and poverty alleviation in another place—Pingyuan Fruit Ranch. Two hundred and four people from 134 households moved from the rocky mountain area to this farm to work under contract. This is the first case of ecological immigration in Guangxi Zhuang Autonomous Region that combined migration placement with poverty alleviation and development with government support.[11]

In addition to migration placements organized by the government, some farmers in mountainous areas moved to contract plots of land through the introduction of friends and relatives. For example, the Chenshuang Livestock Farm near the county seat was a state-owned farm, but after the reform and easing of controls, the farm was dissolved and all 4800 mu land was abandoned. Twenty-four farm households from Shangnan Township in the rocky mountainous area contracted for 1800 mu land in the late 1970s by paying a certain amount of rent to the livestock farm annually. Because they are not migrants organized by the government, their registered permanent residences are still in their original homes. This is not only inconvenient in terms of social management, but psychologically difficult. They live in old houses on the farm, and except for a few families, most residents do not dare to build new houses.[12]

In the early 1990s, after observing poverty in certain areas, the CPC Central Committee made poverty alleviation and economic development a priority, and decided to implement a "poverty-alleviation program" in the "Eighth Five-Year Plan" period. After considering the outstanding achievements of the pilot efforts of Huanjiang Moanan Autonomous County's immigration placement and poverty alleviation and development program, Hechi Prefecture People's Government and the Huanjiang Maonan Autonomous County explored cross-county relocation and poverty alleviation and development. They jointly founded the first cross-county immigration placement farm, Daan Forest & Fruit Ranch that includes 2000 people from 400 migrant households from Duan Yao Autonomous County; the results have been amazing.[13] In March 1993, at the Eighth National People's Congress, Tan Sanchuan, a representative of the National People's Congress and then deputy county chief of Huanjiang Maonan Autonomous County in charge of poverty alleviation, carefully summed up Huanjiang Maonan Autonomous County's successful experience in trying to implement relocation of poverty alleviation and development in the County since 1989, and drafted the proposal, The Implementation of Poor Population Resettlement in Poverty-Stricken Areas, which was submitted to the National People's Congress. As a result, this proposal was approved by the CPC Central Committee and State Council. The CPC Central Committee decided to put the pilot work of "immigration placement of poverty-stricken masses" in Guangxi Zhuang Autonomous Region. Then, Guangxi handed this mission to Hechi Prefecture, and Hechi Prefecture handed it to Huanjiang Maonan Autonomous County. Since then, all policies related to the "immigration placement and poverty alleviation and development" have been introduced and the immigration placement work of Huanjiang Maonan Autonomous County has been conducted as part of an ongoing development period.

By October 2003, a total of twenty-one migration placement sites and sixty-four settlements had been founded in Huanjiang Maonan Autonomous County, and 60,000 mu land had been requisitioned to resettle 6,887 people from 1,454 households from the county. In addition, 207 immigration placement sites for people from other regions had been founded and 320,000 mu land had been requisitioned and founded to resettle 12,575 people from 59,583 households from the counties of Duan, Dahua and Donglan.[14] Accordingly, Huanjiang has become the biggest county of immigration placement and poverty alleviation and development.

A Case of an Immigration Village

In April 2008, I conducted a investigation at Xiyuan Village, Xianan Township, Huanjiang Maonan Autonomous County. This is a small village committee, six kilometers from the township government and 80 kilometers away from the county seat. By March 2008, the total population in the village was 500, with a total of 120 households. All of residents are Maonan. Due to the scattered cultivated lands in Xiyuan Village, the small villages are spread across all the cultivated land. There are about 10 households in the bigger villages, and two or three households in the smaller ones. There is a distance of 0.5-1 kilometers or 2.5-6 kilometers between villages. All the villages are built on the mountainside or the peak near the lands. This part of the rocky mountainous area is more mountains and less lands. The land is barren and there is no river or springhead, and the production and living conditions are quite poor. Even during the relatively long period after the reform and easing of controls, the food ration per capita was only about 125 kg, with less than100kg during bad years. At present, there is no asphalt road in the 4 villages among the 7 natural villages of the village committee, and the village is not yet fully equipped with "three direct links" such as electricity, water, and telephone. In recent years, with the help of institutions like the World Bank, some villagers of Xiyuan Village have built cisterns for rainwater storage, which has partially solved the shortage of drinking water. During the dry season of winter, however, the cisterns only store limited amounts of water, which must be used economically. Generally speaking, the life of Xiyuan villagers is still very poor. In 2007, there were 70 people from 50 households in the village enjoying the minimum living allowance for the rural special poverty-stricken population, which is 14 percent of the total population of the village; the average percentage of people living at the minimum living allowance in Huanjiang County is 4 percent.[15]

For a long time, Xiyuan villagers who wanted to leave the mountains and escape poverty, joined the army, pursued the civil service, or took the university entrance exam, etc. Since the reform and easy of controls, going out to work has become a trend. Using 2007 as an example, 125 farmers in Xiyuan Village left to go work, which accounted for 25 percent of the total population of the village. During the ecological immigration organized by the government, there are more than 20 households that have moved to other areas.[16] During our interviews all the interviewees said that the government has supported Xiyuan Village through the construction of infrastructure, education, anew rural cooperative medical service, and reforestation efforts. However, the natural conditions of Xiyuan Village are difficult and further improvement and development will be hard, so about 30 percent of the interviewees made it clear that they hoped that the Government would help them move or migrate to other regions with better conditions.

All the young people of the Xiyuan Village leave to work outside, and those households that are able also move. This village committee has no schools and great difficulty in getting drinking water and medical treatment. Nearly half of its villages have no electricity or asphalt roads, so it may become a silent village and disappear in the future.

A Case Study of Immigration Placement Farm for Migrants from Other Places of the County

From February 1-2, 2009, I conducted an investigation at the Chengshuang Immigration Placement Farm, Huanjiang Maonan Autonomous County. The Farm is located 4.5 kilometers north of the Huanjiang County seat, and now it is under the jurisdiction of Chengbei Community, Sien Town. In the past, it was the land of the state-owned Chengshuang Livestock Farm with a total area of 4,800 mu. After reform and the easing of controls, the operations of the livestock farm suffered and the land was almost abandoned, so 24 farm households from Shangnan Township in the mountainous area spontaneously contracted for 1,800 mu lands. In 1996, Huanjiang Maonan Autonomous County People's Government requisitioned the remaining 3,000 mu land of livestock farms to settle the poverty-stricken farmers from Shangnan Township (now it is named Shangnan Village) through an agreement. At the end of January in the same year, all 78 workers from 78 poverty-stricken farmer households (one worker per family—mostly Maonan) from the villages of Shangnan, Gedan, Minquan, and Shangdong, Shangnan Township were led by Mr. Qin, a civil servant from Shangnan Township Government to Chenshuang Livestock Farm to carry out migration development.

The first month after they first arrived at the farm, the migrants slept on beds on the ground. After the immigration site was defined, they began to build houses. Besides distributing lands to the migrants, the government also provided 100 Yuan subsidy for living expenses to each person every month in the first year. In addition, the government provided a small subsidy for building materials and purchasing the means of production. Because the land conditions are better than in their hometown, those migrants worked hard and brought their families to farm. The vast majority of migrants have settled on the farm. Those farmers who have moved back to their former homes will be replaced.[17]

A Case of Immigration Placement Farm for Migrants from Other Counties

On February 3, 2009, I conducted an investigation at Shaiguling Yao New Migrant Village, Huanjiang Maonan Autonomous County. The Village lies in the northern part of the county, 4 kilometers from the county seat of Huanjing. Since October 2003, it has been under the jurisdiction of Chengnan Community, Sien Town. In the past, it was a model village for the placement headquarters of Guangdong Province that assists other areas. Construction began on January 1, 1999. The village covers a total area of 5,127.7 mu, and most migrants are from Dachong, Liushan, and Chongshan villages, Longfu Town, Duan Yao Autonomous County and has a total population of 1008 people from 204 households, and all of them are Yao. The lands requisitioned for a migrant farm were formerly waste lands in Beimen Village, Chengnan community, Sien Town and have relatively fertile soil. Since the establishment of the farm, it has received strong support from the governments and relevant departments and institutions in Guangdong and Guangxi, especially Dongguan City, Guangdong Province. A total of 15 million Yuan has been put into the construction of this farm, of which Guangdong has provided 10 million Yuan and Guangxi has provided poverty-alleviation funds of 5 million Yuan. The housing, office buildings, and schools have been built in the migrant village and roads leading to the county seat have been constructed. Wires, cables, and water pipelines have been set laid and ponds, tanks, and dams have been built. About 2,500 mu in waste slopes has been reclaimed for the migrant farm; the migrants mainly plant mulberries, sugar cane, fruit, and other economic crops and grow food crops such as corn, sweet potatoes, and dry grains. Due to adequate policy input, the construction of this migrant farm has been very good.[18]

MAIN MEASURES AND EXPERIENCE OF THE MIGRATION DEVELOPMENT IN HUANJIANG MAONAN AUTONOMOUS COUNTY

Ecological migration began in Huanjiang Maonan Autonomous County in the 1980s. Over the past 20 years, they have insisted on the guiding combined ideology and principles of immigration placement and poverty alleviation and development. Yet, they constantly adjust the specific policies and measures according to actual situations and experiences, so as to make this work develop in a positive direction and provide useful experiences for migration development in China.

Government-led Programs and Policy Guidance

Migration placement and poverty alleviation are complex social system engineering, and its main population is low income individuals who live in remote mountain areas for a long term, with relatively weak economic opportunities, awareness, or the ability to change their living conditions. Due to the different types of work required such as the selection of migration locations, land implementation, planning and building of migration locations, and relocation organization, the completion of large-scale migration is impossible without the strong leadership and organizational implementation of the government.

For example, the land expropriation must be executed through the government's actions. The land expropriation scope of Guangxi mainly includes private plots and surrounding hills which were distributed to local farmers after 1981 but not developed before 1994, the original collective barren hills and wasteland left undeveloped in 1994, and the fields unused for many years.[19] Under the planning of the higher migration authorities, both the out-migration and in-migration governments conducted full review of immigration related issues, including land use. After seeking consent from the population in the in-migration areas, the out-migration government organized representatives who would like to migrate to conduct a spot investigation. After the investigation, whether to carry out the immigration relocation is decided by the population, and the land expropriation procedures are handled by the land bureaus and migrations offices. However, in the course of implementation, there are many difficulties and problems. There is resistance among a considerable number of the original inhabitants in Huanjiang Maonan Autonomous County whose land is to be expropriated. In particular, the problem lies in the acceptance of migrants from other counties. Some people think that, as there is still a large number of rural people in the county who do not have adequate food and clothing, why should they shoulder the "migrant" burden? Thus, the mobilization publicity becomes the most difficult job during migration placement. In order to eliminate the worries of the local population and carry out successful ideological transformation of the population, they carry out legal publicity and education activities about "The Advantages of Migrants" in migration-related villages and towns. They also try to persuade the "die-hard households" in order to make the population voluntarily relinquish their lands. Some people even apply to the government for the expropriation of their lands for their migrant brothers. Some post large welcome banners and send the leaders of their villages and teams to take part in the welcome meeting in the county. Some villages provide migrants with building materials free of charge.[20]

To ensure the migrants at all the migration sites have a relatively stable and harmonious production and living environment, and in order to avoid disputes in the future, there must be distinct boundaries between the expropriated land for houses, mountain forest, agricultural land, etc. for the production and the surrounding rural land and forest in every migration site (point). Therefore, the certificates for land ownership registration and boundary survey maps of land ownership are made for migration sites as required, and bilateral agreements and village regulations with adjacent masses are signed.

Similarly, the patient and painstaking ideological work for the mobilization of migration relocation should also be conducted by the government. Since there are few opportunities for them to interact with the outside world, many people of Zhuang, Yao, Maonan and other minority nationalities in the mountain areas are accustomed to their current living environment and interpersonal relationships, and have great worries about migrating. In this regard, measures have been taken by the relevant government departments. A large number of civil servants are sent by the Huanjiang County Government to the impoverished mountain areas of the county to mobilize the people and organize them to visit and investigate the migration sites. At the same time, they are visited b the now well-off farmers who migrated during the 1950s to 1960s, so they can find hope and see their future. Meanwhile, adhering to voluntary principles, they are permitted to come and go freely. During the early stages of migration, the return rate of migrants was very high. For example, at the beginning of ecological migration organized in Huanjiang Maonan Autonomous County in the spring of 1989, the farmers did not actively apply for it. Only 40 poor farmers in Shangnan Township came to the Pingyuan Fruit Ranch Poverty Alleviation Development Zone as part of the first batch of migrants. Many were disappointed after seeing the land, and five people left at that time and seven left the next day. Finally all of them left the development zone. Later, after repeated propaganda and mobilization by civil servants at all levels of governments, 204 families successively settled in this zone. There is an even higher return rate of migrants in Duan, Donglan, and Dahua counties who moved in Huanjiang Maonan Autonomous County, and a number of migrants even moved in and out several times.[21] As a result, the government announced that land will be dominated by the out-migration migrants for a number of years. In this way, the migrants will have two pieces of land in both the out-migration and in-migration areas for the first several years. In the first few years, some migrants worked both the out-migration and in-migration areas, and the main workforce and even family members went back and forth between the two places. As a general rule, the land in the out-migration area should be taken back by the local collective economic organization after three years of immigration, and distributed to the farmers who have not yet migrated.[22] However, after several years of migration placement, this kind of migration development was widely welcome. Many people actively began to apply for migration, so the return indicators of the population were soon displaced by eligible people.

Government Financial Support for the Construction of Infrastructure

Relocation, long-distance cross-county or cross-city relocation, must have a certain amount of capital investment to address the costs of the relocation process, including transportation, arrangements for the migrant's production and life after relocation, skills training etc. Besides national poverty reduction funds and local financial counterpart funds, there are often national special funds, assistance funds from other provinces and cities, as well as funds or loans from abroad.

Before 1993, funding for Huanjiang Maonan Autonomous County to carry out relocation came mainly from a variety of poverty alleviation funds in the county. After 1993, with the overall launch of migration placement in the region, funding for migration placement in various counties has been coordinated by the Poverty Reduction Program Office of Guangxi Zhuang Autonomous Region. The fixed payment transfer method is applied to the migration funds in the Guangxi Zhuang Autonomous Region, with 3,500 Yuan for an inner-county migrants, 4,000 Yuan for a cross-county migrants in this Region, and 4,500 Yuan for a cross-region migrant. The appropriate regional funding are transferred in accordance with migration planning and the number of successfully resettled people. The subsidies for relocation-related travel by migrants is paid by the out-migration county from development funds. The migration project management costs (3.5% of the project funds) are paid by financial matching funds from Guangxi Zhuang Autonomous Region. From 1993 to 1995, 600 families in Huanjiang were resettled in new locations, and the actual allocated funding was 8.46 million Yuan (65% of paid funds, and 35% of unpaid funds). In accordance with local conditions, Huanjiang Maonan

Autonomous County People's Government related another 120 families, and with difficulty allocated 650,000 Yuan from other financial sources. Unpaid funds are used in a unified manner by the county poverty alleviation company mainly for the land expropriation, infrastructure construction, and business management, of infrastructure construction is the focus. The paid funds are directly to the resettled farmers, who have to borrow and repay the loans. The loan period is three to five years, and may be extended under exceptional circumstances. The loans are mainly a prepaid expense for migrant households. A loan averages 10,000 Yuan is made to each household in advance through installments, according to the production progress and living conditions of the households. [23]

In order to implement the direction issued by the Party Central Committee and the State Council on organizing developed coastal provinces and municipalities to help the poverty-stricken provinces and autonomous regions in western China with the purpose of resolving the food and clothing problem for the rural poor population, the central authorities decided to make Guangdong to support Guangxi. Therefore, the Guangdong Provincial Party Committee and Government contributed 30 million Yuan to support migration and poverty alleviation development in Hechi prefecture. Ten million Yuan of the poverty-alleviation funding was used by Guangdong Province for the construction of Shaiguling Yao New Migrant Village in Huanjiang Maonan Autonomous County.[24] After the freezing rain and snow storms in Southern China in early 2008, most of the houses at Shandong Migration Point in Chenshuang Migration Site were damaged, Guangxi Charity Federation provided 270,000 Yuan donated by Agile Property Holdings Limited of Zhongshan City in Guangdong Province to assist the villagers here to construct new houses.[25]

The starting point for migration placement and poverty alleviation development is to fundamentally improve the living environment of migrants. Therefore, from the beginning, Huanjiang Maonan Autonomous County People's Government has put great importance on the construction of infrastructure for migration sites, transportation, electricity, drinking water, residential community planning, and educational and medical facilities which must be considered and distributed scientifically and rationally, so that the migrants after the relocation can "settle down." From 1993 to 2006, Huanjiang Maonan Autonomous County built 28,058 brick houses and 5,100 square meters of steel-concrete composite farm produce markets for the settlements, 407 human and animal drinking projects, and 407 water tanks with a capacity of 16,280 cubic meters. They also built 610 kilometers of water pipes, 36 kilometers of water channels, 50 small-sized hill pools and reservoirs, 2,785 water tanks with a volume of the 111,200 cubic meters, and 8,600 biogas digesters. Power transmission lines have been built for all migration points so that the migrants can have electricity for production and living. They have constructed 283 village roads with the total length of 428 kilometers, built 63 new teaching sites, and 10,200 square meters of western style school buildings. Primary schools have been set up at all the migration sites.[26]

Mechanisms for Sustainable Development Supported by Scientific Technology

The problems faced by ecological migrants in Guangxi are essentially the concrete manifestation of the contradictions that arise from economic, social, and environmental factors. There are limitations on the production methods and economic structures used by the poor farmers who have lived a long time in the rocky mountainous areas. After the implementation of immigration placement, the areas of arable land the migrants have obtained are limited, so it is impossible to enhance production and increase revenue simply through the expansion of cultivated areas. Therefore, relatocation planning has to pay a lot of attention to the co-ordination of the relationships among population, economy, resources, and environmental factors as well as migration, employment, and development, in order to establish a mechanism for sustainable development. Otherwise, the same mistakes related to human and environment factors will re-occur and new questions will emerge.

Thus, during the implementation process for migrant placement, Huanjiang Maonan Autonomous County People's Government selected production projects in migration sites based on market, technical requirements, land status and other factors, in order to promote industrial development for migrants to ensure that they can escape poverty. When developing migration sites, Huanjiang Maonan Autonomous County people's Government selected industries based on available resources that would support economic development, such as fruit, sugar cane, silkworm, dendrocalamus latiflorus munro, green plum, flue-cured tobacco, and livestock after careful and scientific planning. These efforts have developed to a large scale, beyond their original fragmented plant situation. At the same time, Huanjiang Maonan Autonomous County People's Government attaches great importance on doing a good job in all areas, particularly the product sales services, so as to provide guarantees for the migration and development.

For example, after the establishment of the county's first remote development experimental zone—Pingyuan Fruit Ranch Poverty Alleviation Development Zone—the Sanhua plum was identified as the leading industry and long-term and short-term management policies were formulated. People on the site are required to inter-plant the understory crops at the time of cultivation, manage fruit trees, reclaim lands, and pay close attention to grain production. At the same time, support is given to the persons on site who raise cattle, pigs, chickens, and ducks, so as to increase their income. Because of the proper methods and effective measures, the living standards of the resettled farmers improved significantly the first year. The problem of food and clothing is basically solved the following year. The attention paid to the initial success of Pingyuan Fruit Ranch by the regional and autonomous region government departments provided a reference for the expansion of the pilot and overall launch of the migration program in the whole region.

In order to ensure a good harvest for the migrants and achieve the goal of adequate food and clothing in the same year and growth in a few years,

Huanjiang Maonan Autonomous County People's Government has prioritzed the scientific and technological training of migrants. Together with the introduction and promotion of superior varieties of crops and the constant improvement of the technology of agricultural planting and breeding, the Huanjiang Maonan Autonomous County Poverty Alleviation Office organizes relevant experts to training migrants on the leading industries such as sugar cane, silkworm, green plum, fruit, and livestock. There are flexible and diverse forms of scientific and technological training activities, such as on-site technical consulting, expert technical guidance, and free distribution of scientific and technical materials on the leading industries.

The immigration development of Huanjiang Maonan Autonomous County also benefits from the strong support and participation of the Chinese Academy of Sciences and other units. In 1996, Changsha Institute of Agricultural Modernization of the Chinese Academy of Sciences, Guangxi Regional Science and Technology Commission and Huanjiang County organized the Kehuan Poverty Alleviation and Development Co., Ltd. and set up the Kenfu Remote Poverty Alleviation and Development Science and Technology Demonstration Zone with a total area of 4,000 mu. The construction of the demonstration zone also considered foreign environmental migration experiences and lessons to set up the immigration database and decision support system. It also studied the causes of environmental migration and migration mechanisms, environmental capacity and its constraints, migration modes and their impact on the ecology, society and economy, and then put forward government strategies and optional technology policies, appropriate immigration environmental capacity, and reasonable migration modes. They locally monitor the impact of land-use pattern change caused by migration development activities on soil erosion, soil quality, and ecological security. They collect data and propose reasonable land-use allocation patterns and technical measures for regulation. They also propose and implement the joint-stock and enterprise-patterned poverty alleviation and innovation mechanism of "science and technology institutions + company + base + farmers." This has important significance and scientific value for the implementation of migration and poverty alleviation development in southwest China as well as the establishment and improvement of a theoretical system for environmental migration studies in China. Five years after the implementation of the project, 80 percent of the migrant households in the demonstration zone were using biogas, and 100 percent of the migrants had clean drinking water. Also, a satellite television network was set up and 83 percent of the migrants possessed a television at home. Finally, 20 percent of the migrant households had telephones and more than 30 migrant families had built two-storey buildings. Poverty alleviation industrial development and ecological construction are organically combined. Level terraces and other development and protection measures were adopted for the orchards and dry land, and the forest coverage rate reached 52.7 percent, which avoided the environmental degradation in migration areas. Thanks to the high technological content and innovative operating mechanism, the per capita net income of migrants quadrupled in four years, namely from 390 Yuan in 1996 to 2180 in 2000, which is more than the county's farm per capita income of 2080 Yuan.[27]

The Whole-village Relocation and Management

The eco-migrants in Guangxi generally live in the rocky mountainous areas with few land resources, poor infrastructure, and a poor living environment. Therefore, no matter one's view of poverty alleviation for poor households, the reduction of input costs for poverty alleviation, or ecological protection, the whole-village relocation is a scientific and rational choice. At the same time, since settlements consist of inhabitants from the same village, the original interpersonal relationships can be retained, and the kinship relations of many people facilitate not only the restructuring of community social organization, but also the community management.

Strengthening the construction of the management system is vitally important to the implementation of immigration placement and poverty alleviation. In 1993, an immigration placement and poverty alleviation development leadership group was set up in Huanjiang Maonan Autonomous County. The county magistrate served as the group leader, and the separate control deputy county magistrate, separate control deputy NPC director, and director of the county poverty alleviation office served as the deputy leaders, and responsible persons of other relevant departments served as the members. The group coordinates the migration work of the county. At the same time, the Economic Development Office of the county is an administrative unit commissioned by the county people's government to take charge of the unified leadership and management of the migration sites in the whole county, and it is responsible for the project demonstration, development planning, financing, and objective management of all immigration sites. The Poverty Alleviation Economic Development Corporation of the county is a unit commissioned by the county people's government to take charge of the economic responsibility for financing, management, loan and repayment of the migration funds in the county, and it is responsible for the funding, management, supervision, and recovery as well as the unified management for all migration sites, and they provide preproduction and postproduction services for the technical training, materials, and product acquisitions.

For site department leaders for all migration sites, Huanjiang Maonan Autonomous County People's Government provided issued documents in 1993 that outlined: four to five people should be designated to serve as the leaders for each migration site, including the director, deputy director and peace officer, technician, accountant, as well as teller and keeper; at least three of these people should be the state civil servants. The director must be the deputy secretary of the relocation township or deputy township head. Huanjiang Maonan Autonomous County People's Government outlined the immigration site management system in 1995 and stated that the in-migration township should carry out the administrative jurisdiction to the resettled immigrates in its administrative division, so as to take charge of the population, law and order, education, health, family planning, taxation, statistics, and other administration management and services, and a competent civil servant from the township government should be appointed director of the migration site. It indicated that the out-migration township should be responsible for the organization and relocation of migration placement objects, and a competent civil servant from the township government should be appointed director of the migration site. His main task is to stabilize site members, be responsible for the return of lost members, and take part in the department management. There should be four civil servants at a migration site, namely, the director, deputy director, accountant who is designated by the county poverty alleviation company, and a civil servant selected from the site members, to take the first step toward having site members manage the sites by themselves.[28]

In October 2003, the remote management of the sites was changed to local management, so the management system became entirely consistent with the local one.[29] When the village "two committees" changed their personnel due to the expiration of the office term in 2005, it was mandated that at least one migrant should join the village committee if the village has more migrants in order to ensure that the "two committees" of the village are more representative and are able to carry out the work conveniently. For example, in Jinqiao Village of Daan Township, where migrants account for more than 90 percent of the population, there are three migrants among the five village committee leaders, while in Haodong Village of Minglun Township, there were two migrants among the four village civil servants.[30]

ASSESSMENT AND PROBLEMS OF THE ECOLOGICAL MIGRATION IN HUANJIANG MAONAN AUTONOMOUS COUNTY

At present, the majority of Chinese scholars have acknowledged the necessity of ecological migration in the western region. The main reasons are: (1) in the western region, there is vast land, dry climate, serious soil desertification, and increasingly deteriorated ecological environment; (2) ecological migration is necessary to alleviate the conflict between the quantity and distribution of population and the land bearing capacity; (3) ecological migration is necessary for people to adopt new ideas, change the mode of operation, and realize the urbanization in the western region; and (4) ecological migration is necessary to solve the conflict between the ecological and environmental protection and the enrichment of farmers and herdsmen. In addition, among the measures for the protection of the environment, ecological migration is the one with few costs but more return. Migration development practice has been conducted in Huanjiang Maonan Autonomous County for twenty years. What are its achievements?

Our surveys and other relevant evaluation show that the poverty alleviation based migration development in Huanjiang Maonan Autonomous County has achieved good results in terms of the poverty alleviation or sustainable development and ecological protection, and it has basically realized the expected objectives. These related evaluations are reflected in a lot of project reports and related theses. In our survey, all the interviewed migrants agree that, compared with their previous life, their living conditions have improved, and all their housing, transportation, education, health care, and other material and spiritual life levels have improved. In particular, through the years of practice, the concepts of the emigrated and non-emigrated mountain farmers have changed greatly, and the non-emigrated mountain farmers are greatly shocked by the demonstration effect of migration development, so they hope to join the project.

For example, the migrants of Shaiguling Yao New Migrant Village previously lived on the rocky hill farms with inadequate water, no roads, no electricity and few fields, and the grain (corn) harvested each year only lasts four months. Today, they live in brick rooms and buildings, eat rice, and use tap water. Every household has more than five mu of dendrocalamus latiflorus munro, and some of them plant economic crops such as sugar cane and silkworms. Their family revenues increase year by year, and many farmers now have television sets, telephones, and motorcycles. Their enrichment is secured, and remarkable achievements for the poverty alleviation in other places are realized. According to statistics, the per capita net income of the migrant in the village in 2007 was 2,100 Yuan, and 80 percent of farmers led the well-off life.[31]

The forefathers of Wei Bingkai—an migrant of Yao, lived in a remote mountainous area, twenty kilometers away from the nearest street fair. They lived in a house with a thatch roof and fence walls. They had to go more than twenty kilometers along the mountain paths to fetch water for cooking every day, and the round trip took more than half a day. Corn was the main food for Wei Bingkai's family, but one year's harvest only lasted 10 months and temporary food shortagae occured every May and June. They used to have no rice to cook. He got 8 mu of fields after migrating to Huanjiang Maonan Autonomous County. In 2007, his annual income exceeded more than 30,000 Yuan by mulberry planting and silkworm cultivation as well as sugar cane planting. Two years ago, he built another floor on his building, and purchased the agricultural vehicles and motorcycles in 2008. At the beginning of the introduction of the migration policy by the government, a lot of people did not want to leave their homes, and there Wei Bingkai's two elder brothers were no exception. In the recent two years, his two elder brothers regretted their decision after seeing his increasingly prosperous life.[32]

Since the implementation of migration development in Huanjiang Maonan Autonomous County, the relationship between migrants and original inhabitants has also gradually improved through publicity and other measures. Over the course of history, Huanjiang Maonan Autonomous County has been a place with Maonan, Zhuang, Han, Miao and other nationalities living together, and all nationalities are in harmony with each other. They live in the same areas and accommodate each other, so no serious ethnic conflict or dispute occurred. Such good inter-ethnic atmosphere provided necessary conditions for the implementation of migration placement. Among the nearly 60,000 migrants accepted by Huanjiang Maonan Autonomous County from neighboring counties, most of them are Yao, then Zhuang, with a small number Miao. There are Maonan, Zhuang, Han and other ethnic groups in the surrounding settlements. In the migration placement process, the rationalization of ethnic relations is inseparable for the migration, road building, and expropriation of the mountain areas, and persuasion and coordination of all ethnic groups in surrounding settlements. Although ethnic-related issues and contradictions inevitably occurred during this period, they were properly resolved. The relationship between the migrants and original inhabitants has gradually improved. Such cases were reported by the media. In 2005, relevant research was also conducted by scholars in Huanjiang Maonan Autonomous County. Through questionnaires and interviews, they found that, in Huanjiang Maonan Autonomous County, there were good intentions for relationships between the original inhabitants, between the original inhabitants and migrants, as well as among the migrants. Thus, the distance between the original inhabitants and migrants can be narrowed. At the same time, communication skills of all nationalities in Huanjiang Maonan Autonomous County are strengthened by the tolerant national consciousness and strong inter-ethnic identification consciousness coupled with good intentions of inter-ethnic intercourse of varying degrees, Thus, a bridge for migrants to integrate into the population of all nationalities in Huanjiang Maonan Autonomous County was set up.[33]

Overall, the migration development in Huanjiang Maonan Autonomous County has achieved relatively good results, which have been welcomed and supported by migrants. In particular, the farmers who still live in the rocky mountainous areas have full expectations for the continuous implementation of the migration policy. The above-mentioned interview material on the Xiyuan Village, Xianan Township, Huanjiang Maonan Autonomous County is a typical example.

However, upon affirming the effect of migration development in Huanjiang Maonan Autonomous County, we should not deny that there are a lot of questions, which are mainly reflected as follows:

First, from the migrant point of view, many problems occurred after migrant placement such as social adaptability, inheritance of ethnic cultures, community reorganization, production and life style, ethnic relations, and social security. However, one of the most important is the land system. In accordance with the policies and regulations of Guangxi Zhuang Autonomous Region, after the state acquisition, the land and other fixed assets shall be permanently and collectively owned by the migration placement sites and the land use rights owned by migrants are non-transferable. Part of the reason is that most of the funds for land development are invested by the government, so the land acquired by migrants at no charge becomes a kind of welfare. This prevents the land use rights from transferring, which leads to land abandonment by some migrants for various reasons. For example, there was 1,000 mu of land abandoned in the Shaiguling Yao New Migrant Village in 2008, so the Beimen Villagers whose land was acquired had a lot of complaints.[34] In addition, with the population growth and enhancement of land value after development, serious land disputes have occurred between some of the original inhabitants whose land was acquired and the migrants.

For example, the land of Gedan Village of Chenshuang immigration site was acquired from the Kenmiao Village of Chenshuang Village. The population lived in peace at the beginning, but since the beginning of the 21st century, the villagers of Kenmiao Village have started to prevent the migrants of Gedan Village from cultivating over 100 mu of their acquired land, and cut away the crops in those fields, so these fields were unable to be cultivated by migrants the last two years. Migrants repeatedly went to the Huanjiang Maonan Autonomous County People's Government about this matter, but the problem was still not solved. Therefore, the migrants do not feel at ease, and some people have even moved back to their native homes.[35]

Second, as far as the original inhabitants are concerned, there are also various problems about the changes, community relations, ecological environment changes, and disputes over interests and land. Especially, with the increase in population and expansion of land reclamation areas as well as the acceleration of industrialization and urbanization, the people-land relationship in Huanjiang Maonan Autonomous County has deteriorated.

The reduction of reserve land resources in Huanjiang Maonan Autonomous County not only causes a tense people-land relationship, but also affects the effectiveness of migration development policy to some extent. In addition to the above-mentioned land disputes between original inhabitants and migrants, it also resulted in difficulties on the relocation of poor mountainous farmers in the county. Although Huanjiang Maonan Autonomous County is the largest place in Guangxi for the implementation of migration development projects, among the 66,470 relocated people from 14,029 households, there are 59,583 people from 12,575 households from Duan, Donglan, and Dahua counties, with only a small number of migrants from Huanjiang Maonan Autonomous County.[36] In recent years, no new immigration placement project has been carried out in Huanjiang Maonan Autonomous County, but those poor farmers who live in the rocky mountainous areas are still dreaming of migrating to other regions with better natural conditions.

Third, there is a great difference between the policies for inner-county and cross-county migrants, which results in uneven implementation effects. Before 1993, the funds for Huanjiang Maonan Autonomous County People's Government to carry out inner-county migration mainly came from the county's financial expenditure, so the inner-county migrants only received limited subsidies and migrants were required to invest more human and financial resources in the settlements' infrastructure. At the same time, its process was also a step-by-step one. Therefore, some farmers from mountainous areas cannot take part in relocation projects because of their limited financial resources. However, due to the high attention of the governments at all levels in Guangxi Zhuang Autonomous Region and the strong assistance from Guangdong Province, the infrastructure construction in the settlements for cross-county migrants was carried out rapidly and soundly, and the migrants had a higher level of living. These differences in policies inevitably had a psychological impact on the inner-county migrants.

CONCLUSION

One of the prominent features of the contemporary Chinese eco-immigration policy is to effectively integrate the migration placement with poverty alleviation development, making it serve as an important measure for the Chinese government to implement an anti-poverty plan. Through twenty years of practice, Huanjiang Maonan Autonomous County has become the largest county for migration placement and poverty alleviation development in Guangxi Zhuang Autonomous Region, and the ecological migration works has made outstanding achievements. All of its measures, experience and achievements are of typical significance. On the other hand, some problems have also gradually been exposed and generated. The summary and research on the ecological migration practices combined the migrant placement with poverty alleviation development in Huanjiang Maonan Autonomous County can provide certain reference value for the development of Chinese environmental policies.

NOTES

1. Li Nin, Gong Shijun. "A Discussion on Ecological Immigrant in Ningxia," *Journal of Harbin Institute of Technology* (Social Sciences Edition), No.5, 2003, pp.19-24.
2. Bai Nansheng, and Lu Mai, "Poverty Alleviation Development and Migration: Methods and experiences," *Management World,* No.3, 2000, p.162.
3. Huanjiang Maonan Autonomous County Poverty Alleviation and Development Leadership Group Office, *A Report on the Situation of immigration placement of Rural Poverty Population,* April 13 of 2007, p.2.
4. The materials are from Mr. Qin Yonggui. I interviewed him in February 2, 2009, at the Office of Huanjiang Maonan Autonomous County Poverty Alleviation and Development Leadership Group.
5. Tang Shunlin, Gong Shijie, and Feng Xinbin, "System Dynamics Model for Sustainable Development of Socio-Economic and Ecological Environment in Karst Area-A case study of Northwest Guangxi in China," *Journal of Safety and Environment,* No.3, 2001, p.36.
6. Xu Lianfang, Liu Xinping, Wang Kelin, and tan Hebin, "Research on the Pattern of Land Arrangement and the Countermeasures of Its Security in Immigrant Region of Northwest China—A Case Study of Huanjiang Maonan Autonomous County," *Territory and Natural Resources Study*, No.4, 2003, p.36.
7. The compiling group of Huanjiang Maonan Autonomous County General Situation and the compiling and revising group of Huanjiang Maonan Autonomous County General Situation, *Huanjiang Maonan Autonomous County General Situation,* (Beijing, Minzu Press, 2008), pp.1-3. p.11.
8. Huanjiang Maonan Autonomous County Ethnic Affairs Bureau, *A Basic Report on Huanjiang Maonan Autonomous County,* June 9, 2007, p.1.
9. Huanjiang Maonan Autonomous County Poverty Alleviation and Development Leadership Group Office, *The Review and the Expectation of Poverty Alleviation Action for 20 years*, May 9, 2007, p.14.

10. Xu Lianfang, Yang Chunhua, Kong Xiangli, and Luo Jun, "Land Consolidation Mode and Safety Measures of Southwest Karst Migration Zone—A case of the Northwest Of Guangxi Huanjiang Maonan Autonomous County, " *Geographical Research and Development,* No.1, 2005, pp.66-67.
11. Huanjiang Maonan Autonomous County Poverty Alleviation and Development Leadership Group Office, *A Report on the Situation of immigration placement of Rural Poverty Population,* April 13, 2007, p.1.
12. The materials are from Mr.Zhou. I interviewed him at The Chengshuang Immigration Placement Farm on February 2, 2009. Mr. Zhou migrated to the farm in 1988 from Minquan village which is about 100 kilometers far away.
13. Huanjiang Maonan Autonomous County Poverty Alleviation and Development Leadership Group Office, *A Report on the Situation of immigration placement of Rural Poverty Population,* April 13, 2007, p.1.
14. Huanjiang Maonan Autonomous County Poverty Alleviation and Development Leadership Group Office, *The Review and the Expectation of Poverty Alleviation Action for 20 years,* May 9 of 2007, p.5.
15. I got the statistics at Xiyuan Village Committee in April 11, 2008.
16. I got the statistics at Xiyuan Village Committee in April 11, 2008.
17. The materials are from Mr. Qin Yonggui. I interviewed him in February 2, 2009, at Huanjiang Maonan Autonomous County Poverty Alleviation and Development Leadership Group Office.
18. I gathered the material from my trip to Shuaiguling Yao New Migrant Village in February 3, 2009.
19. Bai Nansheng, and Lu Mai, "Poverty Alleviation Development and Migration: Methods and experiences," *Management World,* No.3, 2000, p.163.
20. The materials are from Mr. Tan, the officer of Huanjiang Maonan Autonomous County Poverty Alleviation and Development Leadership Group Office. I interviewed him at his office in February 4, 2009.
21. The materials are from My field trip to Chenshuang Migration Placement Farm in February 2, 2009 as well as from Mr. Tan, the offier of Huanjiang Maonan Autonomous County Poverty Alleviation and Development Leadership Group Office. I interviewed him at his office in February 4, 2009.
22. Bai Nansheng, and Lu Mai, "Poverty Alleviation Development and Migration: Methods and experiences," *Management World,* No.3, 2000, p.164.
23. Xu Likun, "Investigation Report for Poverty Alleviation Development in Maonan Autonomous County of Huanjiang County, " *Journal of Guangxi University for Nationalities,* Supplement, December 1997, p.121.
24. I gathered the material from my trip to Shuaiguling Yao New Migrant Village in February 3, 2009.
25. I gathered the material from my trip to Chenshuang Imigration Placement Farm in February 1, 2009.
26. Huanjiang Maonan Autonomous County Poverty Alleviation and Development Leadership Group Office, *The Review and the Expectation of Poverty Alleviation Action for 20 years,* May 9, 2007, pp.5-7.
27. Chen Yiyu, " Speech on the Summary and Commendation Conference of 'Sanke' Poverty Alleviation through Sciences and Technologies," (December 2000), www.cas.cn, *Newapaper of CAS "Leaders Speeches."*
28. Xu Likun, " Investigation Report for Poverty Alleviation Development in Maonan Autonomous County of Huanjiang County, " *Journal of Guangxi University for Nationalities,* Supplement, December 1997, pp.120-121.

29. Huanjiang Maonan Autonomous People's Government, *A Report on Poverty Alleviation Development Doring New Period in Huanjiang Maonan Autonomous County,* November 2, 2007, p.5.
30. I gathered the material from Huanjiang Maonan Autonomous County Civil Affairs Bureau in April 10, 2008.
31. Guo Yanqun, "Family-like Close Relations and Harmonious Advance between the Immigrants and Huanjiang Minority Nationalities," *Guangxi News Network—South Morning Post*, December 8, 2008.
32. Guo Yanqun, "Family-like Close Relations and Harmonious Advance between the Immigrants and Huanjiang Minority Nationalities," *Guangxi News Network—South Morning Post*, December 8, 2008.
33. Shang Qiujin, and Su Xianzhong, "On the Poverty Reduction by Emigrating from the Perspective of Ecology of Ethnic Groups," *Journal of Tianshui College of Administration*, No.1, 2008, pp.32-33.
34. The materials are from Mr. Mo Weiheng who is a farmer of Beimen Village, Chengnan Community, Sien Town. I interviewed him in Januray 30, 2009, at my parents' home at the Sien town, Huanjiang Maonan Autonomous County.
35. I gathered the materials from my field trip to Chenshuang Migration Placement Farm in February 1, 2009.
36. Huanjiang Maonan Autonomous County Poverty Alleviation and Development Leadership Group Office, *A Report on the Situation of immigration placement of Rural Poverty Population,* April 13, 2007, p.2.

Index

www.ingramcontent.com/pod-product-compliance
Ingram Content Group UK Ltd.
Pitfield, Milton Keynes, MK11 3LW, UK
UKHW041629031026
15587UKWH00001B/173

9 780739 147429